The Self-Help Handbook
for Small Town Water and Wastewater Systems

by

Jane W. Schautz and Christopher M. Conway

This document has been funded wholly or in part by the United States Environmental Protection Agency (USEPA) under Cooperative Agreement Number CX–821501–01–2 with The Rensselaerville Institute. Document contents do not necessarily reflect the views and policies of the USEPA nor is the Agency responsible for the document's accuracy, adequacy, efficacy, or applicability. Mention of trade names or commercial products does not constitute endorsement or recommendations for use.

Sylvia Bell, Chief
Small Communities Section
Municipal Support Division
Office of Wastewater Management
United States Environmental Protection Agency

Project Officer:
Betty B. Ford

Copyright © 1995 The Rensselaerville Institute
All rights reserved

ISBN 0–9629798–4–8

Library of Congress Catalog Card Number: 94–80130
The Rensselaerville Institute
Rensselaerville, NY 12147
Telephone: 518–797–3783

The Self-Help Handbook
for Small Town Water and Wastewater Systems

CONTENTS

PREFACE   1
Hon. John Dawson,
Arkansas House of Representatives

INTRODUCTION   2

PART ONE: THE STATE ROLE   9
    The Role of Government   11
    What's In It for the States?   12
    Connections in States   14
    The Value of Compacts   15
    Implementation of State Self-Help Programs   18

PART TWO: CRITERIA FOR SELF HELP PROJECTS   21
    Potential   23
    Readiness   28
    Interaction of Potential and Readiness   31

PART THREE: SELF-HELP STRATEGIES AND HOW TO USE THEM   33
    Make All Possible Use of Local Resources   35
    Reassess the Problem and Solution   40
    Practice Conservation   45
    Determine Project Priorities   48
    Choose the Simplest Possible Solution   49
    Involve Local Workers   59
    Borrow or Lease Equipment   69
    Purchase Materials and Equipment Directly   72
    Cooperate with Other Governments   79
    Avoid Duplication   85
    Use Volunteers   87

Become the General Contractor	100
Communicate!	102
Get The Most From the Outside World	106
Legal Help	108
Engineering Help	110
Project Financing	119
Sources of Capital	122
Sources of In-kind Contributions	141
Self-Help as State Match	144
Contractors	145
Insurance	149
Take Essential Actions Before Construction	161
Secure an Engineering Plan	162
Investigate and Comply With Regulations	164
Establish the Right Legal Structure	165
Prepare a Construction Schedule	167
Obtain Civil Service Approvals	170
Plan the Transition from Construction to Operation	173
Prepare for Operation and Maintenance (O&M)	173
Complete All Records and Plans	174
Record Lessons Learned	175
Celebrate!	176
Connect with Other Existing Resources	177
PART FOUR: THE CONTEXT	185
The Need for Improvements	187
The Small Town and the Rural Area	195

PART FIVE: APPENDICES 205

A. Directory of State
 Environmental Agencies A-207

B. Case Study: Western Maryland
 Cooperative Utilities Venture B-221

C. On MONEY... and Knowing How
 Much You Can Afford to Borrow C-239

D. Washington State In-Kind Policy D-243

E. Massena, NY, Request for Proposals E-247

F. Preliminary Engineering Report,
 Water and Sewerage (FmHA) F-275

G. References G-283

CASE STUDIES	
Blanchard and Edison, WA	44
Delmar, MD/DE	46
Marshall, NC	66
New Jerusalem, AR	71
Connelly Springs, NC	88
Dolgeville, NY	123
Heuvelton, NY	133

PART SIX: INDEX 291

Preface

The Self-Help Handbook is for anybody who wants to help people make a tremendous impact on their own lives through improving their own community. We don't have enough money to give people all the things they need, so how can you beat self-help to improve infrastructure?

This book is the Bible of The Small Towns Environment Program (STEP), one of the finest programs that I have ever seen in my life. We are taking advantage of it here in Arkansas, starting several years ago with the New Jerusalem water project right in my own district.

The people in that community got together, got the easements, borrowed a bulldozer from the county, got people to volunteer their time to run it, and they laid the pipe! Instead of having an $180,000 project, they did it for $80,000. Saving money! Saving the taxpayers money!! Also, the cohesiveness that developed in that community was second to none.

Self-help is definitely an idea that all governments — national, state, county and especially local — should support. Many of our rural communities can do what New Jerusalem did. Once people hear of it, the idea will get so contagious that we'll have to give some kind of immunization to keep it from spreading.

John Dawson
State Representative of the 38th District
Arkansas House of Representatives

Introduction

In a Nutshell

This Handbook describes a set of tools that small communities can use to reduce the cost of drinking water and wastewater projects. It is intended as a desktop reference for two primary audiences. One consists of local residents — elected officials, plant operators and concerned citizens — for whom this book provides detailed advice on how to do such a project at a price their community can afford. The other comprises state and federal officials responsible for water and wastewater programs who can use the book for guidance on a new way to facilitate community projects.

Background

The authors are senior staff members of the Small Towns Environment Program (STEP), which is operated by The Rensselaerville Institute (TRI) with major support from the U.S. Environmental Protection Agency (EPA) and The Ford Foundation. TRI, a nonprofit development center established in 1963, works with individuals and organizations to build and test fresh solutions to social, economic and educational problems.

What is Self-Help?

Since 1973 TRI has devoted a substantial part of its efforts to community revitalization. During that time we have developed considerable expertise in the use of techniques that reduce the cost of vital infrastructure projects while increasing community capacity. We call the systematic use of these techniques "self-help."

For our purposes, self-help refers to collective effort: people working together to create or improve a service or facility (for example, a water system) that they will use in common but which is not exclusively owned by any one person or household. It is therefore somewhat different from the concept of "sweat equity," by which an individual or family gains private ownership of something (such as a house) by investing labor in its creation.

Another important distinction involves self-help and voluntarism, which many people think are synonymous. They are not. Self-help as we define it **includes** the use of volunteers as one technique, among many, that can reduce the cost of a needed community improvement.

In brief, with the self-help approach, small communities draw first on their own resources — human, material and financial — to solve local problems. This book tells readers how to apply the strategy to water and wastewater problems.

Why Self-Help?

For state governments, self-help is an effective tool for promoting regulatory compliance. One of its most attractive features is that it complements traditional tools for achieving compliance, such as grants programs and enforcement actions. For local governments, self-help is a way to meet the challenge of creating viable and affordable water and wastewater systems. While it is by no means appropriate for every town, in many situations self-help can bring needed improvements at substantial cost savings. In an era in which the cost of replacing aging infrastructure and complying with environmental mandates is increasingly borne by localities, such cost reduction makes sense to citizens. Indeed, for many communities, there may be no alternative.

The Precedent

TRI's early efforts to fashion and test self-help solutions to critical social problems focused on renewal of dying small towns. During the 1970s we purchased Stump Creek, Pennsylvania, and Corbett, New York, both company-built and -owned settlements. Our renewal projects enabled residents to rejuvenate their communities and purchase the homes they had always rented. In the 1980s, TRI's partnership with the Cherokee Nation provided another example of self-help in action. Relying largely on their own volunteer labor, residents of Bell, Oklahoma — a predominantly Cherokee community in one of America's poorest counties — installed a 16-mile water line.

However, the direct precedent for STEP was TRI's work with the State of New York which resulted in the creation of the Self-Help Support System. Over the last ten years this partnership among TRI and three state agencies (The Department of Environmental Conservation, the Department of Health, and the Department of State) has brought assistance to nearly 150 local water and wastewater projects, with cumulative cost savings of more than $17 million.

The New York experience showed us that, rather than working directly with one community at a time, we could multiply our outreach by helping state governments to play an enabling role vis-a-vis small communities with infrastructure needs. This led to the creation of the Small Towns Environment Program, whose mission is to transfer to interested states the knowledge and skills that TRI has gained after many years of work in the community development field. Through demonstration projects and customized training, STEP helps state staff learn how to facilitate local self-help efforts. To date we have established such a mentoring relationship with nine states, and a number of others have expressed interest in the program.

About This Handbook

The first edition of the Handbook focused on New York State whose Self-Help Support System was the first example of the institutionalization of this approach at a state level. The present edition has been substantially revised and updated so as to be applicable throughout the nation. It contains five parts:

Part One focuses on the state role in self-help. Our contention is that isolated community actions without a clear state enabling role will stop well short of potential. We begin with a look at a new role for state government (the enabler) as counterpart to the traditional roles in environmental protection of provider and enforcer. We then turn to early steps to add a self-help approach to the state's tool kit.

Part Two sets forth the elements that should be present in a community in order for self-help to work. It is important to understand that self-help will **not** work in every community. We have identified a number of factors that are strongly associated with success, and urge all readers to consider them carefully so as to try this approach only where it is most likely to succeed.

Part Three looks at specific strategies for practicing self-help as grouped into five areas. First come local resource questions, where we begin with the difference between adding up the costs and starting with affordability. Next are strategies for connecting to the lawyers, engineers, contractors and other professionals on whom communities must rely for important services, even in self-help projects. Then we discuss essential steps that a community must take prior to construction, followed by advice on making the transition from construction to operation. Finally, we give suggestions on providers of information and technical assistance that can help small systems remain viable.

Part Four situates self-help within a larger context. It includes a discussion (provided largely by our partners of the U.S. EPA) of the nature and depth of the infrastructure problem in small towns, followed by a look at our unit of action — the small human settlement.

Part Five consists of appendices that provide illustrations of key points and tools for broader applicability.

Acknowledgments

The generous support of the U.S. Environmental Protection Agency and The Ford Foundation has made STEP a reality. Gail S. Shaffer, then New York's Secretary of State, was an early believer and has played a crucial role in the success of the Self-Help Support System.

We would like to thank Sylvia Bell, Chief of U.S. EPA's Small Communities Section, as well as our current U.S. EPA Project Officer, Betty B. Ford, and her predecessor, Stephanie von Feck, for their advice and encouragement. Our thanks as well to the many people who reviewed various sections of the book, including Diane Perley of the New York State (NYS) Environmental Facilities Corporation and Doug Ferguson of the NYS Department of Health, both of whom are key members of the NYS Self-

Help Support System; Eric Stockton, who manages the Small Community Outreach Program for North Carolina's Department of Environment, Health and Natural Resources; Bruce Henry, SCORE Coordinator in U.S. EPA Region IV; Peter Shanaghan, Small Systems Coordinator, Office of Groundwater and Drinking Water, U.S. EPA; Beth Hall, Environmental Specialist, U.S. EPA; Nikos Singelis, Program Analyst, U.S. EPA; Kristi Watkins, Mobilization Coordinator, U.S. EPA Region IV; Kermit Holshouser, Mayor of Connelly Springs, NC (site of STEP's first project in that state); Anna Sessum, Community Planning and Development Representative, U.S. Department of Housing and Urban Development; Joni Leithe, Assistant Director, Government Finance Research Center; Elton Balch, Director of Planning, NYS Department of Civil Service; and Maurice Morganstern, Supervising Insurance Examiner, NYS Insurance Department.

We are especially grateful to Alberta Bouck, Christopher Brozek and Kathie Speck of the TRI staff for their invaluable assistance in the production of the new edition of the Handbook.

Finally, we thank our many colleagues in active and emerging STEP states for their energy and commitment. Self-help is not about what you say or even about what you believe. It is about what you do. Thanks, partners, for making it happen.

Harold S. Williams, President
The Rensselaerville Institute
Rensselaerville, NY
January, 1995

Part One

The State Role

The state plays an essential role in enabling self-help. Indeed, the change process begins here.

The Role of Government

In the past, state government, often aided by federal dollars, fulfilled the role of provider for infrastructure needs. The government supplied:

Protection — Safety and health regulations which defined the nature of the problem using external "public good" criteria;

Programs — Generally pre-engineered solutions such as wastewater package treatment plants that would solve the problem;

Money — Seen as the scarcest resource needed to implement a program solution.

The state's behavior then created a corresponding role definition for localities. Communities generally learned to be three things: dependent (since government often liked to deal only with engineers and other "experts" whom its solutions funded); patient (since most expected to rise to the top of the state priority list eventually); and compliant (since the government was presumed to know what the community needed).

These role distinctions for state and local governments simply don't work any longer. Localities must now be helped to solve their own water and wastewater problems. In many cases, there is no other choice. For the state, this is a shift from provider to enabler, bringing a new understanding of the three elements noted as provider functions:

> **Performance targets** in the form of quality control standards that local efforts must achieve;
>
> **Local projects** (instead of external programs) as the solutions;
>
> **Local talent** (instead of money) as the scarcest resource to be found and nurtured.

For communities to take the initiative to solve their own problems, they must move from a feeling of dependence to independence — from patience to impatience. Otherwise, they won't be motivated to act!

What's In It for the States?

Why should a state bother to add a new role to current responsibilities — especially at a time when resources are stretched so thin? The answer is that self-help is not just another new program that competes for scarce resources to provide still another government service. In fact, it is not a new program at all in the sense of heaping still more onto a full plate. Rather, self-help is a new tool to help states deal with their existing agenda.

Self-help has many virtues, including building capacity among small town residents for cost reduction. But its primary purpose is to increase compliance with state and federal clean water requirements. Indeed, under many conditions and for some difficult situations, it can do so at a lower cost than can grant programs or enforcement actions.

Self-help programs, as we define them, are anything but "soft" in terms of dollars in and results out. Indeed, we have developed cost-accounting methodologies not only for the local need in a given self-help project but for state services to support it. When compared to total costs in other approaches, the savings can be very significant. You may want to read a new STEP position paper, "What Does it Cost to Save Money?...The View from the States." It's available for $6.00 postpaid from STEP, The Rensselaerville Institute, Rensselaerville, NY 12147. Phone 518-797-3783; fax 518-797-5270.

Another gain from the self-help approach for state governments is that it can remove some of the pressure. An unfortunate premise of the otherwise laudable pursuit of "reinventing government" is the notion that problems will be solved if government alone changes its ways and wares. Self-help shifts the focus not only to communities but to residents within them. Its premise is that we must "reinvent citizenry" such that people take more personal as well as collective responsibility for solving their problems and maintaining public facilities.

Still another value is that the principles of enabling (in contrast to providing and enforcing) apply broadly to social and human services. In health, for example, experts have long known and demonstrated that preventive behaviors (e.g., not smoking, controlling weight, exercising) are as influential in health as is the entire formal health care system. What is prevention if not self-help? More broadly, governments have begun to question an old phrase: "The Social Service Delivery System." As if social services were a commodity or package that can be delivered in a system to individuals and communities who passively wait for help! Enabling works as well for job creation, housing, and education as it does for public infrastructure.

14 / Connections in States

When Gov. William Donald Schaefer visited the NEEDS project (Neighbors' Endless Efforts Demand Satisfaction) near Hagerstown, MD, he told sparkplugs Sheila Jones (l) and Nancy Russell (r) that the project was amazing.

Connections in States

Water and wastewater systems are often defined as a set of elements (houses and other users, the treatment plant, etc.) that are connected together. In a different way, state systems may be defined by their links as well. These connections are within and among state departments, where walls of turf and responsibility typically encourage strong disconnects.

Within environmental agencies, for example, self-help requires involvement on a variety of fronts. One is the regulatory side, where compliance issues and enforcement actions can help create a sense of an urgent problem. Another is the program side, including construction grants and technical assistance. Still another is the State Revolving Loan Fund (SRF) for wastewater systems (there will be SRFs for drinking water systems, too, if

Congress passes a proposal that is under consideration as this book goes to press).

Whether within departments or among them, mandates and programs for wastewater are often separated from those for drinking water. And wastewater is often divided into separate groups for onsite and central collection systems and needs. While divisions may make functional sense for a state department, they can be barriers for an enabling delivery system that serves customers who wish to view a state agency as seamless — at least for them. This is a primary tenet of the popular reform movement known as total quality management.

These divisions can be bridged in one of two ways. The more difficult is at the structural level, where functions are realigned to reduce barriers. This is complex and often unnecessary. Indeed, a remarkable number of agency reorganizations have far more impact (often negative) on those within the agency than on customers outside of it. A preferred strategy, at least initially, is to handle correction on an interpersonal plane. At the top level, start with a memorandum of understanding or a compact among agency or divisions heads. At an operating level, look for individuals who have, personally, a cooperative bent. They, together, can forge new links that can later be made structural if necessary.

The Value of Compacts

Agreements such as compacts and memoranda of understanding provide a written framework that serves as a blueprint for action. They define not only the hopes and aspirations of the **people** involved but the expected **results**. A compact provides an opportunity for people to work together to define goals, ensure agreement, and produce a written document that can be used to measure progress and keep people on track.

Even though each situation is different, requiring a unique document for a specific project, place and time, it may be helpful to look at compacts developed by others. One of the projects that led to the formation of STEP was a renewal project in the hamlet of Corbett, New York. A compact was developed among residents and staff of The Rensselaerville Institute, and signed in a public ceremony. Its preamble said:

> We give our pledge to rebuild Corbett as a small community in which people help each other... in which we can get a good night's sleep... in which our children can range safely... in which we can feel good about our town, our neighbors, and ourselves... in which we do not waste.
>
> At the same time, we seek a community in which people live and let live, respecting the rights of others to be different. We want people to grow. Some will grow and stay. Others will grow and leave. But for all of us, Corbett will always be home.

That compact went on to define the responsibilities of all persons to make a community renewal project work. Those who signed on took it seriously. While it had no legal force, it proved to have strong emotional meaning. It was taken as a pledge or a promise by most people and was not lightly set aside.

Communities undertaking self-help projects might consider preparing a compact. Here are three virtues of compacting:

- **Compacts are enabling**. Traditional legal agreements are full of qualifiers put in place to protect the parties and to guard against all contingencies. Compacts are agreements based on a common faith that people can and will take action for a broader good. Expectations influence all of us!

- **Compacts have energy**. They gather much more enthusiasm than do the traditional kinds of work plans and charts often used as the only written framework for a project. A key reason: they are very personal.

- **Compacts are results-driven**. At their best they go beyond the traditional "C" words of communication, coordination, and cooperation. They speak to **collaboration**, in which an outcome is created that individual parties could not, on their own, achieve. Collaboration is an expression of interpersonal, not structural, relationships.

Project compacts can usefully include:

- a preamble to set forth a shared vision of what the community should be like;
- a set of principles or beliefs about what is to be done and how;
- a statement of what each party has to do to make the project successful;
- a way to sign on, such that participants make an explicit choice to join in a common framework.

Implementation of the compact may require finesse. In the prototype state for STEP, New York, care was taken not to define the Self-Help Support System in structural terms. No letterhead was printed, for example, since this would force out questions about which Department (Environmental Conservation, Health, or State) should have its name on top. Indeed, this impressive initiative succeeded by being a non-program!

Implementation of State Self-Help Programs

Self-help is not a religion, it's a strategy for change. Therefore, how does STEP help a state move from discussion to action? Answer: energetically and collaboratively!

STEP looks for states that are searching not just to do more with less, but to do it in smarter ways. When STEP is invited to a new state, the host agency is encouraged to include all the other departments and organizations that might have an interest in hearing STEP's presentation of the self-help approach. Attendees are generally from regulatory agencies for water and wastewater plus agencies for community development and economic opportunity. Also usually represented are funding agencies and training centers, as well as nonprofits or academic institutions that offer technical assistance. Allowing for adequate time to respond to all questions, this session may require two to three hours.

The next steps fall along two parallel tracks. The first solicits support and encouragement from top management of the lead agency and explores whether any other agency should be closely involved in a relationship with STEP. Effort then turns to devising a partnership document, a compact (as described in the previous pages) that will specify the separate and shared performance of each signatory. (These are the heads of the lead departments and The Rensselaerville Institute, STEP's parent organization.) The actual execution of the compact is frequently a ceremonial occasion that includes press coverage for the new initiative.

Enthusiasm to get started frequently dictates that both tracks be pursued simultaneously. One purpose of this effort is to identify one or more demonstration projects to test whether self-help is as effective in the new state as it has been elsewhere.

Here again collaboration is crucial. Interested agencies are asked to nominate towns that might be a good fit: towns that have real water or wastewater problems and prefer to be self-reliant. If an unwieldy number of communities are named (Arkansas' first list had 256!), additional criteria of smallness, financial need, public health threat, etc. will eliminate many. Self-help criteria (discussed in Part Two) will reduce the list further, and prequalifying phone calls can lead to a short list: the three to five finalists.

At this point a team of two or three state officials and STEP staff make site visits to explore each finalist's fit. These meetings may be in the town hall primarily with local government officials, in the fire hall with key civic leaders, or in a church or school with a large proportion of the town's residents. The team accepts whatever configuration (and time of day) is comfortable. The agenda includes a presentation of STEP ideas, the community's self-assessment as to its own potential and readiness, and a dialogue about those implications.

The team is eager to learn all it can during each several-hour visit, but primarily it seeks reliable information on two questions: (a) are residents **really** motivated to solve their problem, and (b) is there a local sparkplug to lead and sustain the effort? Affirmative answers to **both** questions are indispensable for a STEP project to be launched.

Immediately upon return to the state capital, the site visitation team reports its recommendations to the lead agency management, and a demonstration project is chosen. (Most states select only one project as a prototype, but the State of Washington chose three!)

The project actually begins as soon as the state team is organized. The latter generally consists of at least one STEP staff person, a decisionmaker who keeps top management informed, and an operations-type person who becomes Project Manager.

Local planning begins at the very next community meeting and includes development of a preliminary timeline, assessment of the town's resources, and selection of the project's core committee.

Since STEP's role is to transfer its concepts, strategies, and techniques to the state's staff, STEP makes no visits to the project site without the knowledge — and usually the company — of at least one other member of the state team. The field site serves as a learning laboratory for the state, an opportunity to participate in the STEP methodology. It is for this reason that STEP does not undertake community involvement on its own; STEP is there to serve as the state's mentor.

Part Two

Criteria for Self-Help Projects

Self-help doesn't work everywhere. It is therefore essential to determine whether prospective community projects are a good fit to the self-help approach. STEP's experience shows that consideration of two groups of factors is required.

The first of these is **potential** — does the community have the **capacity** to make a project work? The second is **readiness** — is the community **willing** to do so at this time? Each cluster of criteria deserves special attention.

Potential

There are five factors of potential that should be present in a self-help community. The first one listed is not only the most important, it's indispensible.

1. "Sparkplugs." Analysis of successful community enterprises usually points to one key ingredient: a person. To a remarkable extent, whether the project is undertaken by a local government, a not-for-profit organization, or just a collection of motivated residents, we find that success is not explained by the availability of money, severity of the problem, or adequacy of the plan. Success is determined by a person who is able to take an idea and make it work. We call this person a "sparkplug."

This kind of person is generally characterized as an entrepreneur, one who starts things. Entrepreneurs have long been celebrated as the most critical ingredient in new venture starts and expansions that feature innovation rather than business as usual. We have found that the same kind of entrepreneurial characteristics noted in individuals in business also pertain to community sparkplugs. Such leaders are increasingly thought to be a necessary component in any effort to innovate with limited funds.

Let's consider for a moment how investors identify entrepreneurs. Venture capital firms (those who invest in new businesses) are composed of a specific kind of investor who finds and bets on entrepreneurs. Unlike banks and most other institutional investors, they are not looking for an assured rate of return on their money. Rather, they become part owner of new or expanding businesses that appear to have high promise. For those enterprises which do "take off," the venture capital firm may get twenty — or more — dollars back for every dollar invested. At the same time, for those enterprises that falter or fail, the investment must often be written off entirely.

Every year venture capitalists are presented with hundreds of would-be entrepreneurs, each fervently believing that his or her idea will make millions of dollars in short order. These investors have developed a substantial body of research to help them define and select entrepreneurs.

Most studies are empirical rather than theoretical. They begin by putting successful entrepreneurs in one pile and other individuals in another and asking, "What's the difference?" Among approximately twenty attributes usually generated by studies of entrepreneurs, there are six distinguishing characteristics that show up consistently. Our experience has shown that individuals who are successful in sparking community projects tend to have these same six qualities.

- Sparkplugs have **energy**. They work long, hard, and with enthusiasm. They are driven by a need to accomplish, to try things out, to achieve results. In a community project, that energy is not only an example, it's contagious.

- Sparkplugs take **personal responsibility**. They tend to believe that they are responsible for what happens and do not blame or credit fate, luck, or chance. Too many people excuse failure by citing lack of government grants, bad weather or something else beyond their control. A sparkplug looks for ways around barriers, not accepting "no" for an answer.

- Sparkplugs tend to have a **narrow and intense focus**. They are not out to attack multiple problems simultaneously. Their objective is to complete one highly specific and concrete undertaking.

- Sparkplugs are **tenacious and resilient**. The strength of their vision of the successful outcome propels them through temporary setbacks and insulates them from discouragement. Their optimism is based not on a mindless Pollyannaism, but on an unshakable belief that there **is** a way to solve the problem.

- Sparkplugs are excellent **learners**. Indeed, in the for-profit realm, many venture capitalists would rather invest in an entrepreneur who has failed than one who has not, other things being equal! The reason? Sparkplugs are good at learning both from their own experience and that of others. They don't insist on reinventing the wheel — a common mistake of many newly designated community leaders.

- Sparkplugs take **moderate — not high — risks**. He or she has the determination as well as willingness to go a bit out on a limb to try a new way when the old way doesn't work. The status quo may be the **cause** of the problem; a new approach may be precisely what is needed. Like entrepreneurs in business, sparkplugs circumvent traditional thinking to find new routes to old destinations. They focus more on opportunities than problems. Note that successful action does **not** require extreme risk, just a moderate, informed bet on the entrepreneur's tested skills and other known resources.

Sparkplugs are not necessarily the best managers, nor are they always the most articulate critics. Rather, they are those rare people who combine optimism, talent and motivation to conceive of a better way to do something and, more important, put their ideas into practice.

How can a potential sparkplug be identified? Generally, by his/her itch to start new things. It could be new businesses, new clubs and groups, new and novel vacation plans, or virtually anything else. Entrepreneurs may well **not** have been involved in many past community activities. Indeed, many may come solely out of a business background where they have found new venturing supported to a greater extent.

Some community-based sparkplugs may feel uncomfortable being compared to entrepreneurs in business. They should not. While no analogy is perfect, the point here is that people who start things and see them through to fruition tend to have similar attitudes and behavioral traits whether the venture be motivated by community spirit or the desire for profit. Clearly, sparkplugs in both business and community development are persons with the disposition and ability to turn ideas into reality, to lead change by example. In both cases, we have individuals who "push the envelope" to bring innovation to people whose needs are not being met by existing goods and services. It is in this sense that the person who leads a volunteer effort to install a water line because there is no money for paid staff can be compared to someone who founds a business to offer a better way to deliver packages overnight.

2. Past experience with successful self-help projects. For people as well as communities, the best predictor of how they will behave in the future is the way they have behaved in the past. If a local group has built a fire hall, playground, or church, the lessons they learned can well be transferred to a new activity. Even if the leadership is different, participants will be energized by the memory that the community was successful previously.

3. Community cohesiveness. If the community has two or more factions that have a history of distrusting each other, community-wide commitments are less likely to succeed. This is not to say the community must be homogeneous. In fact, some of the best projects come from communities with a great deal of diversity in such things as income, age, outlook, and educational level. But communities that are divided into "camps" by geography, length of residence, income, etc. — or show a history of **becoming** divided in past times of crisis — are not apt to work together effectively for a self-help project.

4. Demonstrated capacity in skills needed for the project. Here the question is not that residents of a community are generally "capable," but rather whether they have the **specific skills** which will be needed. Is there a bookkeeper here who will help? Does anybody have experience with construction? How about somebody who knows about securing and recording easements?

5. A critical mass of residents, at least 15 households within the target area. To realize the greatest cost savings and to finance the improvement, it's important that there be enough people involved to spread the costs to an affordable level. It's also important that the distance between connections not be excessive.

Readiness

These five practical indicators of potential are only half of what it takes. The other half is desire. Does the community have the will to proceed? Here are eight key factors:

1. A strong local perception of a problem. If only the state regulatory agency understands that a given community has a drinking water or sewer problem, then the responsibility for solving it may be perceived as belonging to outsiders. Only when those directly affected come to believe strongly that **they** have a problem will they be motivated to solve it. While outside agencies can encourage recognition of a problem, they really can't "sell" this kind of readiness. It has to come from within the community.

The village well in Connelly Springs, NC before the water sources dried up. This problem eventually created the motivation for the state's first self-help project.

Local leaders are in a far better position to raise awareness about the problem. They can do this in any number of ways. One is to provide compelling examples to show how bad things can get if the community waits too long. Bottles filled with murky water, photos of septage pooled on lawns, and similar visual evidence can serve this purpose. We know of a few leaders who have even suggested temporarily **inducing** worse conditions to dramatize the situation. This step may not be appropriate or even ethical, but the point is clear: when people feel threatened by a crisis, they will take action.

2. A perception that local implementation is the best and maybe only solution. While self-help efforts often lead to unexpected social and psychological benefits, sometimes the most convincing argument for a self-help project is that there is no other way. Most people would prefer to have others do the job if monies were available, but these days Americans may have to come to grips with tighter budgets. The rediscovery of local capacity frequently begins only when people understand clearly that they have no choice but to rediscover it.

3. Community confidence that local residents can do the job adequately. In addition to seeing a problem and recognizing that self-help is the necessary solution, residents need to have confidence that the community can make it happen. Indeed, studies show that both groups and individuals must believe that they can be successful before they will even try. Often people aren't even willing to **identify** the problem until they have a sense that a solution is out there somewhere just waiting to be found.

4. Support of local government. It is certainly true that private citizens alone have accomplished public improvements. It is also true that a resistant local government makes it more difficult to bring off a successful self-help infrastructure project. The municipality is the legal entity that must sponsor and probably finance the improvement, and is one obvious source of labor, equipment, and administration. For those unincorporated areas that meet our other criteria, the county usually assumes formal responsibility. The fact that a self-help project will bring credit and prestige is usually not lost on politicians.

5. No competing priorities. Both the local government and residents must agree that solving this particular problem has more urgency than any other local concern. Without such a single-minded focus, energies become diluted. Self-help projects require more effort, not less, to produce a satisfactory conclusion; one self-help project at a time is as much as most small communities can handle.

6. Some previous assessment of the problem. Chances are improved if some work has already been done to analyze the nature and extent of the difficulty. This might have been a formal study or even informal but serious discussion. It takes time for concepts to mature and consensus to be reached, so mobilization for action can happen more quickly if the need has already been defined and both technical and financial options explored.

7. Both public and private willingness to pay increased costs. While self-help can certainly lower costs, it cannot eliminate them. A good starting point is to make sure that those who will receive the new benefit are willing to increase their share of costs for the service if that will be required. STEP starts by asking how much people can afford and then seeks to contain costs at that level. This approach motivates the locality and the state to find creative ways to deliver a first class project at a discount price.

8. Enthusiastic and capable support of the community from the county, regional office of the lead department, or other sub-state agency. Considerable sources of assistance to small communities exist within the state government. In STEP states, authorization for that help has already been given through the compact — and empathetic, technically qualified individuals are identified as collaborators. STEP-state project managers can broker services to provide professional advice for all aspects of the project from preplanning through operation and maintenance. They are contacts for "one-stop shopping."

Interaction of Potential and Readiness

While most communities and agencies are clear about avoiding risks when potential and readiness are low, many times mistakes result from proceeding when **either** potential or readiness is high. **Both** elements must be present!

Once the possibility is raised for the undertaking of a needed improvement, community interest and involvement will build very quickly. That momentum can be sustained only if realistic appraisal demonstrates that the project is both desirable and feasible.

The interaction of readiness and potential is paralleled by another interaction that may be important, the relationship between enforcement and readiness. It is our sense that enforcement can often be a critical part of the enabling role of state government in promoting readiness. Indeed, there may well be no other way to convince people that they have a problem than to initiate the strongest possible enforcement procedures. In some instances, local government may wish to actually work with the state to speed up the action timetable as a way of putting the self-help effort into gear. In addition to the threat of fines and penalties, another sanction can work well in a growth area: a provision that there can be no new connections to a water or wastewater system until the present system is brought up to standard.

Part Three

Self-Help Strategies and How to Use Them

MAKE ALL POSSIBLE USE OF LOCAL RESOURCES

The first questions to be asked have to do with what a community can do for itself. In general, self-help strategies are meant to achieve what economists call "import substitution." Simply put, you and your neighbors are doing for yourselves some things that you might otherwise have paid outsiders to do.

You may find STEP's "Flow Chart for Project Managers" helpful in understanding what's involved. The chart is a simplified diagram of a typical small town water or wastewater project, showing the interrelationships of tasks and milestones from the very beginning to the very end. This is a generic work plan that serves as a checklist to make sure that all aspects are anticipated and addressed.

Each of the twelve strategies discussed in Part Three can bring substantial savings. In order to evaluate which ones to use, however, you must learn the comparative cost savings of each approach. We find that this may vary considerably. In one project, a community may save 20% of the cost by purchasing materials directly and 5% by using municipal employees. In another project, the most significant cost savings will be a lower interest rate on a loan.

However, quality is never to be sacrificed. STEP urges scrutiny of every single line item of the budget, asking the question, "How can we get equal or **better** performance for less or no money?"

We can begin by looking at a very general model of costs to anticipate the range of savings from each category of self-help that we will explore in this manual. To begin, let's take a hypothetical $216,000 water or wastewater project done in a "retail" way, using a consulting engineer and general contractor. Here is one cost scenario for such a project:

A. Direct Materials Cost (competitive retail bid)	$ 70,000.00
B. Materials Markup (@ 5% of A — to cover bidding, acquisition, handling and inventory)	3,500.00
C. Direct Labor (subcontracted competitive bid)	70,000.00
D. Engineering (design & certification @ 10% of A, C, & F)	15,435.00
E. Engineering inspection (@ 10% of C)	7,000.00
F. Contingency Costs (@ 10% of A, B & C)	14,350.00
G. Overhead and Administrative (@ 10% of A through F)	18,028.50
H. Profit (@ 10% of A through F)	18,028.50
I. TOTAL:	$216,342.00

While the proportions of labor and materials will vary, as will the exact labels for other costs, the ratio shown here of direct to indirect costs is well within typical ranges. And we do **not** suggest that they are unreasonable. Just as the direct cost to make a refrigerator or a car is no more than half the final sales price, the direct costs of a water or wastewater project are far from the total bill. In the above scenario, for example, contractors' markup for materials is justified by their services in materials acquisition, handling, movement, and, perhaps, interim financing. Their markup for labor and equipment is justified by the anguish of hiring and dealing with subcontractors (who may not show up when they say they will!).

So too, is the "overhead and administrative" item real money. One cost, for example, is marketing. Think of the trips that are made (usually by the highest-paid people) to present proposals to town meetings. Someone has to pay for that time — especially when the presentation does not lead to a contract. And, of course, the company needs profit to stay alive.

Our intention is not to cast stones at this retail approach. Rather, we are offering an **alternative** when that approach is unaffordable.

Take the above scenario and assume a strong self-help stance. Here's what might happen.

1. The town bargains for an engineer at a flat rate of $50 per hour with an "upset" or cap of $7,500 for design and supervision.

2. The town buys its own materials for $65,000 and absorbs the handling costs within its ongoing operations.

3. The town hires subcontractors for 50% of the work and supervises them using a public works superintendent already on the public payroll. The other 50% is done by municipal employees (a mix of long-term and temporary people) who work on the new project during slack periods. (Note that some states have provisions against hiring temporary employees as a way to avoid paying a contractor's higher wages.)

4. No "overhead and administrative" or "profit" lines are added to the project; overhead and administration can be part of municipal functions, and the municipality doesn't charge a profit.

The total costs:		
	Engineering	$ 7,500
	Materials	65,000
	Labor	35,000
	TOTAL:	$107,500
	SAVINGS:	$ 97,100 or 45%

But that's not the end of the savings. Let's now include a piece to be discussed in a later section of this book — access to money. If we assume that the projected "retail" cost of $216,342 was paid at 7% from a bond program over 15 years, the interest cost alone would total $139,956.

Now, take the lower base of $107,500. Suppose you have only $2,500 in cash, but have bargained effectively with the local bank, pointing out that the village or town has substantial deposits from time to time and that the bank has a number of home mortgages for properties in the project area. As a result, they give you a **10-year** commitment for $105,000 at **6%** interest. This results in total interest payments of only **$37,661.**

Let's bring that down to the user. Assume 200 connections (households) on the system that must pay off the improvement (principal and interest) through a rise in their monthly service charge. Under the retail approach, the monthly increase per user is $9.90; under the self-help approach, the increase is only $5.95. And in this example, you pay for **10** years instead of **15.**

The savings can be substantial. But there is no magic wand. Self-help involves sweat, some risk, and a strong testing of a community's spirit and capacity.

To get a fix on your local situation, a good starting point is to establish a "retail" cost baseline for the proposed project. What would the prices be for materials, labor, and needed equipment using traditional contracts? If you have an engineer's plan, it will often detail the costs by category at prevailing retail rates. However, Diane Perley of New York's Environmental Facilities Corporation reminds us that some engineers lump together costs such as legal, administrative, interest, etc. If the engineer's estimate is not sufficiently detailed, you may need to request a specific breakdown of costs. Without such a complete explanation, assumptions (such as local wage rates, whether the attorney is already on retainer, State Revolving Fund (SRF) vs. bank interest costs, etc.) cannot be checked and adjusted.

In addition, a number of tables showing "industry standards" are available to help with your computations. When you then calculate reduced costs for doing it yourselves, you will have a clear sense of savings and a useful technique to apply to other local undertakings.

REASSESS THE PROBLEM AND SOLUTION

You may think you know what has to be done to solve your community's water or wastewater problem. The solution may be obvious, or it may have been presented to you by experts. Nevertheless, it's always prudent to stop and ask if you've really found the **best** solution to the **real** problem. The strategy you've been told to use may be far more expensive than it has to be, or it may simply not address the root cause.

This sort of mismatch between problem and solution is one of the biggest ways that communities waste money. In the good old days when a great deal of money was available from federal and state coffers, communities and their agents frequently saw no reason to question Cadillac solutions or the professionals who proposed them. Now that funding is harder to come by, it is necessary to re-examine all options.

The following examples are drawn from the work of the New York State Self-Help Support System (SHSS). They illustrate the importance of reassessing problems and solutions before you to start to build. (Note: In addition to the city, New York has three other levels of incorporated local government: the village, the town and the county.)

* * * * * * * *

Vernon, NY. Residents of this Village of 1300 people were paying unusually high rates for the water supplied to them from an adjacent community. The Village was investigating wells as a new source when SHSS staff proposed a different solution: why not simply bypass the current supplier and build a new line to the original source? This suggestion led to a partnership between the Village, the Town of Vernon, the City of Oneida, and a citizens' group from the Town. Compared to the original estimates for developing a groundwater supply, costs were reduced by half a million dollars, and the annual savings for water alone ($64,000) will pay for capital construction costs over a ten-year period! In addition, the Village mayor and the Town supervisor negotiated with their local bank for a loan at 4% interest. The Village provided oversight for all engineering and construction, agreeing to supply water to previously unserved Town residents in the area where the line passed through Town property; in exchange for this service, the Town waived the taxes it could have levied.

* * * * * * * *

Ripley, NY. The Town of Ripley needed to comply with the recent federal regulation requiring filtration of water drawn from surface sources. However, the initial price tag was $1.6 million with annual user fees of $1000 per household! A new Town supervisor had enough faith in both his municipal employees and an alternative technology (slow sand filtration) to take a chance on the self-help approach. The construction of a water filtration facility was a major undertaking, and skepticism was rampant. But with the exception of a few tasks that were contracted out, Town employees built their own slow sand plant, thus saving local citizens $1 million in capital costs and at least $70,000 per year in operating costs.

* * * * * * * *

Speculator, NY. The Village's water came directly from a nearby lake, chlorination being the only treatment. The existing chlorine injection point was only a few hundred feet from the Village, and thus did not provide sufficient contact time for proper disinfection. A consultant proposed to improve the disinfection process and reduce taste and odor problems by building a ground-elevated storage tank and relocating the chlorination facility to the site of the tank. The estimated capital cost was $500,000.

The Village board sought advice from SHSS staff, who recommended against the consultant's plan for three main reasons: (1)

the proposal would not solve the disinfection problem since it would still allow the chlorinated water to mix with raw water in the transmission line; (2) continuing to use the lake as the water source would eventually necessitate costly filtration technology as the federal Surface Water Treatment Rule was soon to go into effect; and (3) if the Village found a groundwater supply, the proposed storage tank might turn out to be in the wrong location.

With the help of the SHSS, Speculator hired a hydrogeologist who was able to locate a suitable groundwater source. Once this new source was found it became clear that the proposed storage tank would indeed have been in the wrong place. Thus, in addition to the savings it achieved through self-help techniques ($600,000), the Village was able to avoid buying a costly white elephant.

* * * * * * * *

Prospect, NY. The Village had recently developed a groundwater supply, and the mayor was making arrangements to purchase a used water tower to provide for storage needs. At one dollar, the price seemed like a bargain, but SHSS staff pointed out that erecting and painting the elevated tank would be very expensive. On their recommendation, Prospect instead purchased and installed a new ground elevated tank, which met the community's needs while saving about $50,000 in capital outlays. And by selecting a glass-lined tank that requires little upkeep, the Village will save as much as $200,000 in maintenance costs over the life of the facility.

* * * * * * * *

There are examples of this strategy from other states, as well. On the following page is a case involving two small communities in Washington.

When doing your reassessment, remember to look at the problem with a long-term view, asking some basic questions about the community's future. On the issue of growth, for example, do you want to encourage or limit it? If you desire more people and industry, then the smallest acceptable diameter of pipe may not be a wise choice when, for a slightly higher price, you can provide larger pipe to serve many more users. If you wish to preserve the present size and scale of your town, the smaller pipe will help to accomplish that goal.

CASE STUDY:

BLANCHARD AND EDISON EXAMINE PROBLEM AND SOLUTION

Blanchard and Edison are tiny settlements within three miles of each other, and very close to Puget Sound. (Edison has 200 people in 78 homes and Blanchard has 60 people in 22 homes.) Their shallow soils, very high water table, and pre-1940s building standards (which applied to 87% of their homes) contributed to the widespread failure of onsite systems there. This situation endangered public health and threatened to shut down the local shellfish industry, a significant source of employment for area residents.

Despite the county's urging them to form a sewer district, get a grant and be done with it, the communities recognized their need to educate themselves both on the causes of their problem and options for solution. They contacted every organization they could think of to request information, including Washington State's Department of Ecology; Island County Health Department; Skagit County Planning Department; Region 10, U.S. EPA; EDASC; Big Lake Sewer District; Washington State Sewer & Water Association; Puget Sound Water Quality Authority; Stack, Chambers & Porter, consultants; IMCO General Construction; and the Samish Watershed Management Committee.

The designation of Blanchard and Edison as a STEP community reinforced their determination to do as much as possible themselves. Their research continued: a survey conducted by the county sanitarian and volunteers led to dye tests of every household, revealing failure of more than 50% of the systems.

Residents built monitoring ports to measure wastewater and tidal activity in both wet and dry seasons. They even got the Burlington Northern Railroad to provide a crane and operator for two-and-a-half weeks to assist local volunteers in cleaning out plugged drainage ditches. This latter action not only resulted in the first dry ditches residents had seen in years but also eliminated mosquitoes!

Community effort produced solid information on which new systems could be designed. Blanchard learned that while they needed more expensive solutions than they had previously, properly-designed onsite systems would still work there. At press time, Edison is still researching options. For both, however, local decision to learn for themselves will result in savings of tens of thousands — possibly hundreds of thousands — of dollars.

For more information, contact: Doris Robbins
583 Ewing Court
Edison, WA 98232

PRACTICE CONSERVATION

Conservation is one of the cheapest ways both to eliminate the need for costly construction and to reduce maintenance costs. By being more careful with water usage, a community may even be able to prolong the life of a marginal water or wastewater system. There are two major approaches to conservation: **behavioral changes** and **technology**. Behavioral change makes a difference in your lifestyle, as in watering the lawn on odd or even days but not both. Technology brings some sort of mechanical solution such as faucet aerators and low-flush toilets. The most effective conservation practices incorporate elements of each. For behavioral changes to work, you really need some sort of feedback, some way of letting people know that they have done well or need to do better. In Stump Creek, Pennsylvania, the community installed a large gauge on their standpipe water tank so that everyone could see how their usage was affecting the supply. In Corbett, New York, if the reservoir is down, the chief water operator literally calls residents and urges conservation.

A more straightforward technique should not be overlooked: water meters. If the rate schedule is realistic, users pay according to their consumption — thereby rewarding those who use less. Meters should be a part of any new water or wastewater project; in fact, meters **must** be installed if the project is funded by Rural Economic and Community Development (RECD), a new federal office which now handles the water and sewer grant/loan programs formerly offered under the Rural Development Administration (RDA), and before that by the Farmers Home Administration (FmHA). As one federal official observed, "We don't buy gasoline from an open hose. Why should we do so with water?" And once the meter rates are set, they should be subject to periodic review to fine-tune the system as it evolves.

CASE STUDY:

DELMAR PRACTICES CONSERVATION

Delmar straddles the border of Maryland and Delaware and is one of only six towns in the entire country that cross state boundaries. The community was plagued by a large unexplained demand on its water system. Not only was the water being wasted, but increased pumping caused the electric bill to skyrocket. The archaic system had no meters to identify leakage, and there was also a question of fairness. The flat rate for their many senior citizens was probably higher than usage required.

With STEP's assistance, Delmar undertook a mandatory meters program in which homeowners could choose either to pay $200 and have their meter pits dug by the town, or to pay $100 and dig the 3' x 5' pits themselves. Local civic organizations, churches and individuals did the digging for those who were unable. This large amount of contributed labor saved the system nearly $400,000!

The project's chief challenge was its organization. In addition to the local sparkplug and his management team, there were also block captains to make sure that every pit was properly prepared when the city crews came through to do the hookup.

Delmar's Water Department staff went out of their way to provide technical support. They distributed lists of what to do — and not do — and publicized a telephone hot line residents could call even nights and weekends to get information or to request onsite advice.

The Delmar project demonstrated both self-help and conservation through the town's ability to correct the leaks and residents' newfound motivation to limit their water consumption. The project was also a breakthrough for Maryland's Department of the Environment; MDE accepted the value of the self-help work as local match against state funds for the later rehabbing of the whole system.

                Contact: Roberta Ernest, Town Manager
                          Town of Delmar
                          100 South Pennsylvania Avenue
                          Delmar, Maryland 21875
                          Phone: (302) 846-2664; (410) 896-2777

Roberta Ernest, Town Manager of Delmar, MD and DE. Delmar's motto is: "The little town too big for just one state."

Conservation can be a particularly important element in managing onsite wastewater systems. In places where there are less-than-optimal soils and a high water table, the expected life of septic systems might be considerably less than the design period of 15 to 20 years. Typical designs assume a household size of 3.5 people and a consumption rate of 100 gallons of water per person per day; however, the actual usage can be reduced to 50 or 60 gallons per person per day — or even less. In comparison to the standard of 5 gallons per flush that we knew until the 1980s and 3.5 gallons that succeeded it, the new nationwide standard is 1.6 gallons per flush for residential toilets. Even so, there are toilets available that use as little as **one quart** at a time! Low-flow faucets and shower heads can also significantly reduce water consumption and extend the life of septic systems.

Water management principles include the environmental as well as financial benefits in more efficient use of the water you have rather than paying to provide a greater supply. Water is becoming too expensive to be squandered in inefficient waste transportation systems. Economic incentives can spur conservation: those who are frugal should pay less.

DETERMINE PROJECT PRIORITIES

The very thought of undertaking a large and complex improvement can frustrate local residents into paralysis. Too often, communities assume that their financial resources and resident capacity are just not enough to cover all of the anticipated costs. In such circumstances it is very helpful to "chunk down" or phase the improvement into manageable pieces that can be approached individually and in a timely sequence. The simple action of dividing and downscaling the improvement into separate phases often will let you begin now!

If only a neighborhood or a particular section needs to be done immediately, perhaps that much could be undertaken. In addition, the smaller job would provide training and cultivation of skills that could be reemployed to advantage on subsequent portions.

The Village of Boonville in Oneida County, New York, has operated on this principle for some time. Their revolving five-year plan calls for capital improvements to occur each construction season with financing handled on a year-by-year basis. As they complete each year's set of tasks, they extend their schedule one more year beyond present limits, thus accomplishing steady replacement and upgrading.

With such an approach, expectations can be met. Responsible budgeting is encouraged as residents become accustomed to the need to provide for improvements and pay their fair share.

CHOOSE THE SIMPLEST POSSIBLE SOLUTION

Your state regulatory agency may have long required that the town's engineer examine a variety of alternatives, giving advantages, disadvantages, and costs for each one. This investigation may be necessary for even preliminary regulatory review. However, it is fair to say that, in the past, not all alternative solutions were investigated wholeheartedly. Because many engineers, public officials, and the general public were most familiar with mechanical, high-tech answers, less attention tended to be given to lower-tech options.

In the glory days of high grants, design overkill was quite common. One reason was that engineers, contractors, and everyone else involved with the system faced reduced liability if the system were so overdesigned that small failures would be absorbed. Secondly, a more elaborate system allowed engineers and contractors to charge higher fees since the engineering and construction are so much more complex. Finally, when engineers can choose a predesigned solution (such as a package treatment plant), they sometimes charge the same fees even though the system design has been done by the vendor and previously approved within the state. All of these factors can work against the selection of a simpler solution even when the latter makes more sense.

Working with the engineer, project leaders should carefully compare the various technical options in terms of the initial capital cost, annual operating and maintenance (O&M) costs, the expected life of the improvement, and the replacement cost. Here are some additional considerations to keep in mind when developing alternatives:

- **Plan realistically.** Small, low-cost infrastructure may permit affordable housing to be built for young families or individuals who would otherwise leave. In communities where open land is available close to town, relatively simple water or wastewater systems can accommodate the demands of commercial businesses and manufacturing.

- **What is the least complicated, least expensive system that will meet community needs?** Most operation and maintenance costs are fixed, whether the plant is running at half or full capacity. Customers of an under-utilized system assume the costly burden of paying for potential service that might never be needed. Similarly, highly sophisticated systems are energy-intensive, chemical-intensive, and labor-intensive, but might offer only marginal improvements in water quality over simpler methods.

The experience of New Berlin, NY, offers an example of the benefits of choosing less complex solutions. The spring that supplied the Village was sometimes affected by surface runoff that caused the turbidity of the water to exceed allowable levels. Local officials were considering developing a new well which would have cost $200,000 or more. Self-Help Support System staff suggested a simpler and cheaper alternative: isolating the veins by installing precast concrete rings with gravel packing. The Village opted for this solution and rehabilitated the well at a cost of only $20,000.

Communities may be surprised to learn how many choices they actually have. The following pages are excerpted from a U.S. EPA pamphlet called, "Small Wastewater Systems: Alternative Systems for Small Communities and Rural Areas," (Publication No. 830/F-92/001, May 1992). These examples are included to stimulate questions rather than provide answers. Your state wastewater agency may be able to provide a list of installations that use some of these techniques so that field trips can be arranged for inspection and discussion of performance.

1 Septic Tank & Gravel Absorption Trench

This is the most common system used on level land with adequate soil depth above the water table. Heavy solids in the liquid settle and greases float to the top of the tank. Bacteria break down some solids. The liquid flows from the tank through a closed pipe into perforated pipe and into gravel-filled trenches where it seeps into the soil. Bacteria and oxygen purify the liquid as it slowly moves through the soil. Inspection ports permit checking liquid depth. Regular pumping of the tank reduces the solids discharged into the trenches and extends the life of the system. Using two compartment septic tanks and resting the trenches (#4) are also recommended to extend trench life.

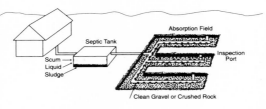

2 Septic Tank With Serial Distribution

Starting with the highest, each trench fills completely, then overflows through one drop box to the next. The effluent floods all soil surfaces. The drop box enables inspection of the system and control of discharge into each trench. Capping the pipe outlets in the upper trench forces resting. Serial distribution automatically loads upper trenches and minimizes the loading on lower trenches. Used on gently to steeply sloped sites.

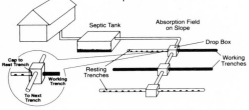

3 Septic Tank & Leaching Chambers

Open bottom concrete chambers or arched plastic chambers create an underground cavern that stores effluent. The effluent floods the soil surface prior to seeping vertically through the bottom of the chamber.

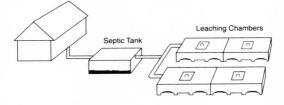

4 Septic Tank With Alternating Trenches

One set of trenches rests while the other treats the liquid from the septic tank. This design extends system life and provides a backup should one field clog. For system repairs, a new field and valve box may be added to the old system. The new field works while the old field rests and renews. Switch the fields annually in the summer.

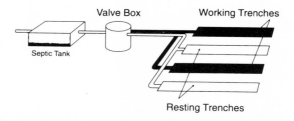

5 Pressure Dosed Distribution

A pump or siphon doses a pressure distribution manifold that disperses the effluent evenly to each trench. Dosing prolongs system life by flooding a larger area and by forcing the exchange of air in the soil. Dosed systems are more common for larger flows. The pressure manifold can include valves or plugs that permit more control over trench loading or trench resting. Annual inspection is suggested.

6 Shallow Trench Low-Pressure Pipe Distribution

Small diameter pipe, located at a more shallow depth than a conventional system, receives pumped effluent. Effluent moves under pressure through small holes in the pipe and soaks the entire trench network area. Even dosing of more open and aerobic soil horizons improves treatment. Used in areas with high groundwater or shallow soils (because it places the treatment higher in the soil profile) or on steep slopes that require hand excavation. Professional maintenance is needed to flush the lines annually.

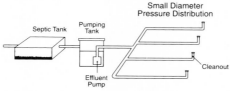

7 Pretreatment & Soil Absorption

Pretreatment addresses the need to treat higher strength waste (such as from restaurants) and can help repair biologically overloaded systems where no additional absorption area is available. Aerobic treatment systems and filters can be used for this purpose. For aerobic treatment (called "package plants"), wastewater and air mix in a tank. Bacteria grow in the tank and break down the waste. For filters, septic tank effluent passes over porous media that trap the solids. Bacteria that grow in the media break down the waste. Professional maintenance by certified operators and a lot of energy are required for aerobic systems.

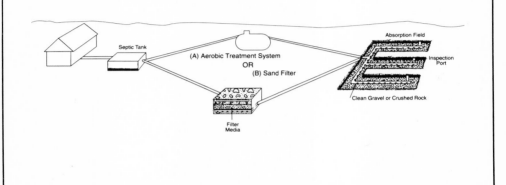

8 Septic Tank & Mound System

Pumps dose effluent (#6) into a gravel bed or trenches on top of a bed of sand. Sandy soil carefully placed above the plowed ground surface treats the effluent before it moves into the natural soil. The system extends onsite system use in areas with high groundwater, high bedrock, or tighter clay soils. Regular inspection of the pumps and controls and flushing of the distribution network are needed.

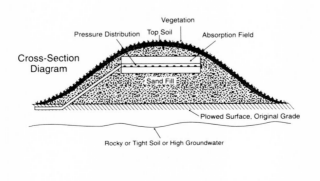

9 Evaporation & Absorption Bed

Effluent from a septic tank or aerobic tank flows into gravel trenches or chambers in a mound of sandy soil. Less permeable soil placed at the surface of the mound helps shed rain from the system. Trees that grow around the system and plants on top of the system pull liquid from the sand and transpire the water into the air. Some effluent may seep into the soil. This system requires a climate where evaporation consistently exceeds rainfall.

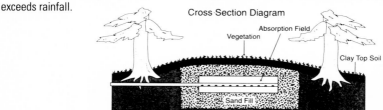

Cross-Section Diagram

10 Septic Tank, Sand Filters, Disinfection & Discharge

Open or buried beds of sand may receive single or repeated applications of effluent. Effluent passes through the media and drains from the gravel and pipe network below the filter. Effluent may be discharged to the environment directly or into a soil absorption or land treatment system (#16). Disinfection often precedes discharge into a stream or land irrigation. Certain types of filters can significantly reduce nitrogen and may be used in areas where soil absorption is not possible. Requires inspection and periodic maintenance. Surface discharge requires management.

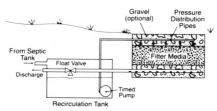

11 Constructed Wetlands

Effluent from a series of septic tanks passes through a bed of rocks planted with reeds. Liquid evaporates and drains into a soil absorption system or discharges. Used for additional treatment or where soils are not suitable for absorption. Discharge usually requires disinfection.

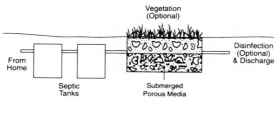

12 Holding Tank

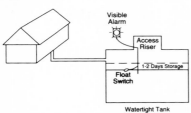

Sewage flows from low-flush toilets and water-saving fixtures into a large watertight storage tank. The alarm in the tank signals the owner to have the sewage hauled away. Only recreational housing utilizes holding tanks because of the high hauling cost. Public management is frequently required. Contracting for hauling helps to reduce costs.

13 Lagoon

A series of septic tanks or other pretreatment systems (#7,#10,#11) discharge into a lagoon. Sunlight and long storage times support the natural breakdown of the waste and die off of harmful organisms. Effluent evaporates, slowly seeps into the soil, or receives further treatment through land application (#16). Onsite lagoons require large lots and may be fenced.

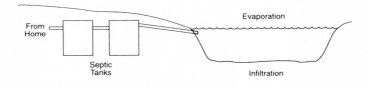

14 Waterless or Ultra Low-Flush Toilet System

Composting Toilets: No Water
   Serve commercial and single family units. Well-designed units produce a dry mixture that should be managed by professionals. Reduces discharge of nutrients into water resources. Electric vent, fan, and heating element common. Proper care is essential.

Incinerating: No water
   Electricity, gas, or oil burns solids and evaporates the liquid, which is vented to the roof. Small amounts of ash are removed weekly. Proper care is essential. Limited to less frequent use sites, such as recreational cabins.

Water Conservation Toilets: Low water
   Low-flush toilets use 1.6 gallons or less per flush. They generally cost slightly more than conventional units, but pay for themselves by lowering the water bill. They perform well. Many work as well as 4 gallons per flush models.

Recycling Water: Low water
   Treated wastewater or graywater recycles to flush toilets. Treatment systems use electricity and require professional maintenance.

15 Dual Systems

Two systems treat the waste. Composting toilets or low-flush (1.6 gallons or less) toilets coupled with a holding tank (#14,#12) exclude nutrient rich toilet wastes (blackwater) from the wastewater disposal system. All other household wastewater (graywater) must be treated in an approved septic tank and absorption system, which is usually smaller.

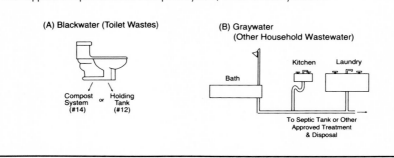

16 Land Application

Effluent from a septic tank is further treated (#7,#10,#11,#13) and stored. Timed sprinklers apply the effluent at night or below the soil surface to plants and trees in a large treatment area. Protects high groundwater in more permeable soils as plants take up nutrients and water. Disinfection and fencing may be required for individual home use. More common in warm climates, but not widely permitted by health authorities.

(A) Slow-Rate Land Treatment

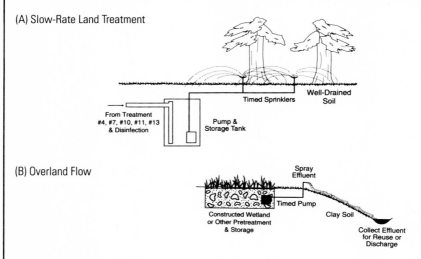

(B) Overland Flow

Treated effluent from a lagoon (#13) or wetland (#11) is sprayed on the surface of a gentle, grass covered slope. Effluent flows over the clay soil through the grass and collects at the base where it is disinfected before being discharged. Best for tight soils where absorption systems are not possible. A professional operator usually cares for the grass and disinfection system.

17 Small Diameter Gravity Sewers

Liquid from a septic tank flows under low pressure in 3-inch or larger collection pipes. Houses below the pipe must use small pumps (septic tank effluent pumps such as #19A and #20). Houses higher than the pipes may drain by gravity. Larger developments favor treatment by a discharging technology such as #10, #11, #13, or #16. Common in rural areas where the community treatment site is generally downhill. Central management is required.

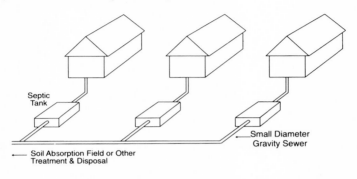

18 Vacuum Sewers

A vacuum station maintains a vacuum in the collection lines. When the sewage from one or several homes fills the storage pit, a valve opens, and the sewage and air rush into the collection line toward the vacuum station. Pumps in the vacuum station transfer the sewage to a treatment system. Power is required only at the vacuum station. Most economical where many homes are served or in areas with high excavation costs and lift stations. Requires a professional operator.

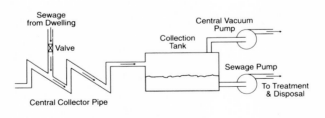

19 Pressure Sewers: Grinder Pump (GP) or Septic Tank Effluent Pump (STEP)

Sewage is first pretreated in a septic tank or grinder pump and then a pump forces the liquid through small diameter lines to a conventional gravity sewer or to a neighborhood treatment plant such as #10, #11, #13, or #16. The community usually owns and operates shared pumping units. Plastic lines located near the surface ease installation and reduce cost. Best for low-density or slow-growth areas or where conventional sewers are costly. Central management is required.

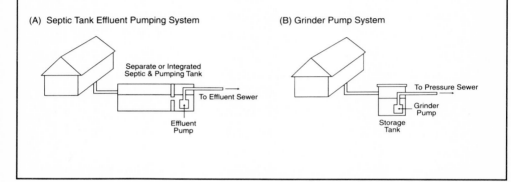

(A) Septic Tank Effluent Pumping System

(B) Grinder Pump System

20 Alternative Effluent Collection System

Liquid from most onsite septic tanks flows by gravity in small diameter effluent lines (#17) to a small neighborhood pump station on public property. A few homes below the sewer may also use small effluent pumps. The neighborhood lift station stores the liquid then pumps it into a higher pressure sewer going to a treatment system. This design can cut costs in flat terrain or where one pump unit can easily serve a number of homes. Central management is required.

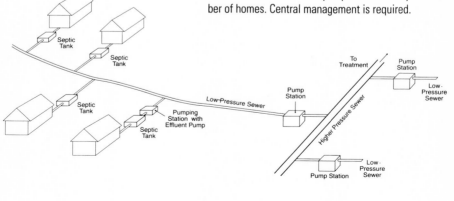

INVOLVE LOCAL WORKERS

Force Accounting. Through the use of municipal employees as project labor, often called "force accounting," communities can save substantial amounts of money. Labor costs may comprise as much as 60% of total project expenditures, so anything that can save money in this category is worth considering. Some of the obvious tasks for local employees on a water or sewer project include the following: surveys of property owners; taking the many photographs that are needed before, during and after construction; clearing and grubbing rights-of-way; making signs and directing traffic; and restoration of lawns and shrubbery. Although inspection is a crucial part of construction, it can also be handled by a qualified person in your vicinity such as a retired engineer in collaboration with the design engineer.

For municipal workers, the wages, hours, and conditions of employment are **fixed by the locality,** not outside agencies or legislation. In general, prevailing wage rates or union rates apply to all personnel hired through outside contracts, but **not** to those people on the municipality's payroll. However, if a **contractor** employs his own labor, he may be required to pay the prevailing wage rate, depending on state laws.

Force accounting is recognized and permitted by U.S. EPA and most states even in those projects that receive government funds, but strict adherence to accounting and record-keeping procedures is required. Funding agencies may ask applicant communities to supply information on the following items:

- local labor and equipment rates;
- ability to accomplish the work;
- accountability procedures;
- procurement of materials;
- insurance and liability implications for the owner;

- guarantees/warranties on self-installed material;
- civil service or state wage rate procedures.

One possible objection to force accounting has to do with what regular work is being omitted during the project's construction. To determine the impact of interruption, communities might calculate dollar values for each set of tasks. How does the regular work compare in short-term and long-term benefits? And how long will the diversion continue? In general, special projects and ongoing work can be reconciled through flexible scheduling... and prevention.

Force accounting may be easier to bring off in cases where the water or sewer district boundaries are coterminous with the municipality. We have heard another objection in some towns that it isn't fair for workers paid from taxes of the whole town to be creating a benefit for a particular, localized area. It **is** fair provided that equipment or services provided by the town are **reimbursed** by the water or sewer district, so one place **is not** taking advantage of the others.

Cautions. As those who have done it will tell you, one of the most important considerations in a successful force account project is the determination of where responsibility lies. Or, to put it more simply, who makes the decision if something unexpected happens? No one can anticipate exactly what field conditions will be — or even precisely what materials may be needed. For first timers, the biggest challenge is the ability to deal with the unexpected. This can be difficult due to lack of experience and, perhaps, incomplete knowledge of available resources.

Communities using force accounting should be clear on these issues whether or not an outside agency demands it. Here are some additional recommendations:

Use of local labor and equipment
- Predetermined rates for labor and equipment, as well as methods of accountability and reimbursement, should be established and agreed to in writing by employees, contractors and vendors before construction begins.

- There may be statutory limits, as in North Carolina. Also, whether municipal employees are permanent or temporarily hired for a particular project, they may need to be approved by the state's Civil Service program. (See Page 170 for more information). Of course, if present employees have job descriptions that fit the new project, no further notification to Civil Service may be necessary.

Ability to accomplish work
- To ensure a successful project, the normal procedure is to have complete project construction documents, hire an experienced and well-qualified contractor and have onsite inspection to guarantee contractor compliance with construction documents. Self-help projects must provide similar competence and vigilance.

- Do local labor forces have sufficient experience in the work required? What else have they done that demonstrates their skills?

- If a project coordinator must be hired, what is that person's experience? (Check out applicants' resumes.)

- The project engineer should be present on the site at critical times of construction as well as for testing and startup.

Accountability Procedures
- You can use procedures adopted by state and federal agencies as a model for force accounting.

- A time sheet should be completed daily by individual employees with a description of work accomplished and the number of hours charged to the project. The time sheet should be signed by both the employee and the project coordinator.

- Equipment utilization should also be recorded on a daily basis with a description of work accomplished and number of hours charged. This report, too, should be signed by the principal operator and the project coordinator.

Reimbursement Rates for Labor and Equipment
- Labor reimbursement rates are the actual hourly rates paid to the individual by the town or village, plus an additional amount to cover the municipality's costs for benefits and insurance.

- Hourly charges for equipment can be based on current rates established by the state.

- Additional equipment required for the project can be rented from local contractors, rental firms or individuals. It's a good idea to get several quotes or use bidding procedures to determine fair compensation.

- Recognize that if the project is to occur in an unincorporated area, a larger unit must become responsible for legal and financial commitments. The water or sewer district must then reimburse the larger unit (city, county, etc.) if the larger unit is to provide services to the district.

Except for rural water or sewer projects funded by Rural Economic and Community Development, the federal **Davis-Bacon Act** requires that workers be paid the prevailing wage if **all** of the following elements are present:

- the contract is in excess of $2,000;

- the United States is a party to the contract or the project will receive federal funding;

- the contract involves the construction, alteration, or repair of public buildings or public works (although the Act may also apply when local governments use federal funds to assist privately-owned facilities that serve the public, which is permissible in certain circumstances); and

- the contract involves the employment of mechanics and/or laborers.

Of course, if the project in question will not receive federal monies, Davis-Bacon does **not** apply. However, state wage rates may apply for some projects and civil service procedures must still be followed for municipal employees.

The Equal Employment Opportunity Act prohibits discrimination with respect to compensation, terms, conditions, or privileges of employment because of an individual's color, religion, sex, or national origin. The act applies to municipalities, but is only triggered when **all of** the following elements are present:

- an employer relationship either exists or is contemplated, and

- the employer is engaged in an industry affecting commerce, and employs at least 15 employees for each working day in each of 20 or more calendar weeks in the current or preceding calendar year. The act does not apply to government contractors, volunteers, or retired people.

Some states require a conscientious effort to involve **Minority Business Enterprises** (MBEs) and **Women's Business Enter-**

prises (WBEs) if your municipality has received state assistance. The federal government usually requires such an effort if the project gets federal funds. However, be aware that implementation procedures may vary among the various federal agencies. Some specify numerical goals for W/MBE participation, others require a report as to whether such firms have been involved in a project, and others don't even ask.

To learn how the regulations apply to a particular project, contact the funding program directly. Note that the EPA most often participates through a state vehicle such as the State Revolving Fund, and very seldom offers financial assistance that bypasses the state.

In most states, municipal workers MUST be covered by **Workers' Compensation**, although this protection is available only to employees, not to contractors (considered to be consultants). In cases where the distinction is not readily apparent, factors such as the following are taken into account:

- The amount of supervision: does this person set his/her own pace and operate independently?

- The way s/he reports: does s/he report daily, weekly, or occasionally?

- The method of payment: is s/he paid in a lump sum or at fairly long intervals — or is the pay period regular and frequent?

- Provision of equipment: does s/he supply his/her own tools?

- Certification: is there evidence of outside affirmation of competence, such as a license?

Not having access to Workers' Compensation, a consultant may sue the municipality if there is an injury. This is one good reason for the community to be adequately insured. (See pages 149-160 for other considerations regarding insurance.)

A great many communities have used their own people for capital improvements; some will tell you that's the only way they could have gotten their projects done. Of the many STEP projects that have used local labor, Marshall, North Carolina is a typical example (see next page).

CASE STUDY:

MARSHALL USES MUNICIPAL WORKERS TO GET WATER

Marshall is a community of about 800 people located in the western mountains of North Carolina. Its water supply problems began in the mid-1970s, when the dam for the town's storage facility was found to be unsafe. Forced to abandon its traditional source, the community started using groundwater. Within a few years, however, the water from the municipal wells declined in both quantity and quality. To make matters worse, the town needed to replace the old cover on its one-million gallon reservoir.

With assistance from STEP and the state's new Small Community Self-Help Program, Marshall devised an ambitious plan to supplement the town's water supply and replace the old reservoir cover using municipal labor and equipment for much of the work.

Town crews connected a productive well to the water mains some 3000 feet away and built a structure to house chlorination equipment. To protect water quality, municipal workers removed the reservoir cover, drained and cleaned the reservoir, and prepared the site for an aluminum geodesic dome (installed by a contractor).

The "retail" cost estimate for the project was $227,000. By using its own workers, Marshall not only achieved direct savings of $78,000 (34%), it leveraged the necessary outside financing. Grants from Farmers Home Administration ($110,000) and the Appalachian Regional Commission ($77,400) required a 20% local match, but the town didn't have that kind of money. However, the federal agencies agreed to accept Marshall's in-kind contribution in lieu of cash (this was the first time such a financing arrangement was used in North Carolina). Detailed calculations revealed that the value of the labor to be contributed by municipal employees was over $47,000. This meant that the town only had to put up $189 in cash! The self-help savings also meant that Marshall needed only half of the Appalachian Regional Commission funds and could return the rest.

This case shows that small communities with infrastructure needs don't have to accept the role of passive supplicant of advice and money. By contributing other municipal resources instead of funds it didn't have, Marshall showed its willingness to play a direct role in solving its own problems.

Contact: Frank Shelton
P.O. Box 548
Marshall, NC 28753
tel: 704-649-3710

Other Sources of Labor. Somewhere between hiring your own work force and using local volunteers is contracting with a neighboring municipality to perform the work. Even another town's pay scale is cheaper than prevailing wages and may present only the challenge of scheduling. There are even more sources of personnel you might also check out. These include programs of your state or county social welfare agency, state and local departments of corrections, and local nonprofit organizations.

In many states, employable recipients of public assistance or "welfare" may be assigned to work on a municipal project. Regulations vary, but here are some examples of the kinds of restrictions that may be placed on the use of such workers in public projects:

- The persons assigned must not be replacing regular employees or those union members who might normally be hired.

- The workers must not be given tasks beyond their physical capabilities. On the other hand, if the project will give them an opportunity to learn a skill while on the job, the local agency may be even more eager to cooperate.

- Participants must be insured, be protected by adequate standards of health and safety, and receive the same coffee breaks and lunch periods given other employees. The local social services agency may provide minimum Workers' Compensation, but often the user entity must carry liability insurance as well as supplemental Workers' Compensation.

- Hours may relate directly to the amount of participants' relief payment calculated at minimum wage or comparable wages for other employees.

- In some cases the local social services department will provide a small stipend to cover the cost of lunch and transportation to and from the work site. Note that there may be restrictions as to how far the worker can be required to travel.

- Project officials will undoubtedly be expected to keep good records on the participation of such workers, and may be subject to specific reporting requirements.

A phone call to your local social service agency may be the most practical first step, after which a written request could follow.

State and local agencies that oversee correctional facilities often have programs that allow certain inmates to work on public projects. Eligibility may be limited to inmates convicted of non-violent offenses or persons who are otherwise considered to pose no threat to the community. If you think this source of labor would be appropriate for your project, contact the agencies in your area to learn what programs are offered and what requirements the employer must meet. Also check with local courts to see if those given community service sentences can be assigned to the project.

BORROW OR LEASE EQUIPMENT

Just because your jurisdiction doesn't own all of the equipment you may need, it may not be necessary to buy or even rent it at high prices. It is quite likely that a neighboring town has the equipment or item you need. And with early discussions and suitable arrangements, they may be pleased to lend or lease it to you. Scheduling is crucial here, of course, so that all parties may be served without severe inconvenience to anyone.

The search should not be limited to local governments, either. You should also approach local businesses and organizations to ask what they could spare for a brief time. You may be surprised at their willingness to help for a public cause.

Self-help projects make maximum use of whatever local resources are available. In Connelly Springs, NC, this included farm tractors and hay wagons loaned by local residents.

Cautions. Before any equipment leaves its home garage there needs to be a clear understanding about liability, especially if this arrangement is not a part of an already-existing cooperative agreement. Two basic questions need to be addressed, preferably on paper in the form of a contract between the owning and using entities: a) liability, and b) insurance. In many cases, it is wise to also pay at least a token rental (e.g., $1.00) so as to place the transaction in contractual terms. You should discuss your plans with both your attorney and insurance carrier just to make sure. It won't improve local relationships to have to deal with these matters **after** someone is hurt or the equipment breaks down.

If you can't get contributed or borrowed equipment, investigate both renting and leasing. Compared to outright purchase, leasing may require smaller monthly payments; it also eliminates the need to find a buyer a few years hence. Leasing companies may offer more favorable terms than banks, preserving the municipality's borrowing capacity.

Venard Norful, Sparkplug of the New Jerusalem, AR, project and also President of the local fire company.

CASE STUDY:

NEW JERUSALEM BORROWS EQUIPMENT

Being from an unincorporated hamlet of only 150 people didn't stop the residents of New Jerusalem, Arkansas from bringing clean, abundant water to the community. They were determined to overcome their longstanding problems with water so impure it wouldn't even get laundry clean. That's when they even had water, which was frequently not the case in the summer.

The New Jerusalem effort, STEP's first demonstration project in Arkansas, was headed by sparkplug Venard Norful. Residents were not only motivated, they had the skills needed to extend water lines for the 4.5 miles throughout their community.

The project would not have been possible, however, were it not for Ouachita County's willingness to lend New Jerusalem necessary equipment and services.

Use of the county's backhoe was particularly appreciated. Because officials in the County Department of Public Works recognized the credentials of at least five community operators, the county backhoe was made available most Thursday afternoons for weekend work.

In addition, the county loaned trucks to help with backfilling trenches, and then later with spreading gravel over driveway cuts.

The community found these and other strategies allowed them to spend less than $80,000 on a project estimated at $180,000, a reduction of over 55%!

                Contact: Mr. Venard Norful
                                926 Ouachita 51
                                Louann, Arkansas 71751

PURCHASE MATERIALS AND EQUIPMENT DIRECTLY

Any municipality may make purchases directly from suppliers and most already do this for regularly used items. The self-help approach assumes that the same strategy used to purchase typing paper and office machines can also work for buying a huge pile of pipe or cinder blocks. The item purchased is different, but the same rules apply.

Start by asking nearby contractors whether they maintain an inventory of the materials you need. Since there is a high cost for contractors to carry excess materials (storage space, the cost of idle capital, and perhaps insurance), they may be happy to help you out at a discount if you explain your awareness of their situation and offer to replace their costs with cash.

You might also inquire from other municipalities, especially if your system is attempting to tie into their larger system. They might be persuaded on the basis of quality assurance and compatibility of their own inventory.

Also, do not ignore the possibility of what a contractor in Elkhart, Illinois calls "scratch and dent" materials: items that may have slight surface flaws that don't affect their working parts or even damages that can be eliminated. This gentleman reasons that as some short lengths of pipe are required on most jobs, why not cut them from a joint whose defect is confined to one end rather than waste the whole thing? Obviously such purchases must be made with utmost care so that the overall quality of the job is in no way compromised. And, of course, the salvage operation shouldn't cost more in labor than it saves.

In addition there are established aftermarket companies that offer recycled materials for resale. These are very few and not well known. One such company is SOS in Tampa, Florida (tel. 813-621-5801), offering substantial discounts on hard-to-find used pumps, fittings and atypical supplies. They operate on a bid basis — you give them a list of what you need and they will bid on providing it. SOS has been in business since 1985 and guarantees their materials for one year. They also do something manufacturers do not do — issue full credit and pay freight both ways if the material is not appropriate.

However, Doug Ferguson, P.E., of New York State's Department of Health urges communities to consider carefully the questions of liability and warranties. If the pipe leaks after the community bought it but a contractor installed it, who is responsible? Should the community be buying these materials at all?

Bidding and Alternatives. In many states, municipal law requires that contracts for public works and large purchases be advertised and awarded to the lowest bidder. The bidding process (preparation time, advertising costs, opening and review of bids, etc.) can add a fair amount to the cost of a local project. You should find out how much this might be. Usually this requirement is triggered only if the amount of money involved exceeds some threshold; contracts valued at smaller amounts do not have to be put out for bid. While it is illegal in some states, in others it is often possible to divide a project into smaller pieces, **each one** with a price tag below the established threshold — and therefore not subject to public bidding requirements. You may be able to break materials into different types to avoid bidding, but examine your state laws carefully to make sure your action is appropriate. In addition, the law may provide for exceptions to the general bidding rules, such as when a public emergency necessitates immediate action.

For purchases that do not exceed the legal bidding threshold (and the figure to use is the **engineer's estimate**, which is not necessarily the final cost), municipalities should do comparison shopping before acting. Find out which vendors in your area offer the best prices, service and guarantees. Be aware that one may be better than the others about stocking odd fittings. Since you can't be sure that your original order won't have to be revised due to unexpected developments, it will save time and, therefore, money to know in advance where you can go for unusual parts.

Another way to avoid the costs of bidding and possibly get a real cost break is by buying through the state contract if your state follows this procedure. Due to the huge volume of its purchases, a state government can get discounts that would never be offered to a comparatively small-scale buyer. Check to see if your state publishes a catalog to facilitate local government participation.

In some other states, there is agreement that the biggest cities' negotiation establishes the price for all other municipalities. It's worth investigating whatever is possible in your state, since the price savings alone can reach as much as 20% — to say nothing of the additional 5% to 10% saved by avoiding the bidding process. Another advantage of this strategy is that a municipality is less likely to have difficulty if state auditors review the transaction. Purchases made through other vendors tend to draw far closer scrutiny — especially if bidding is involved.

Municipalities can use the state contract not only for buying, but also as a guide to their own purchasing elsewhere. Unique market conditions may favor another source at a particular time.

Do weigh the cost savings against the disadvantages, however. Investigate all conditions of the purchase. First of all, is the item you need even available through the state contract? If it is, but deliveries are made only to a distant point, might the transportation costs exceed money saved?

Even when bidding is required, realize that this doesn't necessarily lead to the lowest prices. The process imposes costs on vendors of goods and services, too. Studies of water systems, wastewater systems, and the construction of public buildings show that the private sector could often do the job for much less money if local governments didn't have to solicit bids. Many firms simply won't deal with government because of added constraints and delayed payments, so perhaps the lowest-margin contractor isn't available to you. Those who bid have to add charges for preparation time, filling out forms, inflation, etc. Further, some firms assume that when dealing with a government all workers must be paid the prevailing wage rate (usually union scale) and all products must be union made. As a matter of local practice or policy, this may or may not be true.

When bidding is required, it need not be done passively. Employ what Lynn Palmer, Executive Director of the Washington County (MD) Sanitary District, refers to as "active bid solicitation." Here are some techniques:

Use your own people to the maximum extent rather than relying on an outside purchasing agent. Develop your own resource pool.
- Prepare lists of all vendors derived from your own knowledge, recommendations of others, *Dodge Reports, American City and County Municipal Index, Public Works Manual, Utility Supply of America Blue Book,* state lists, and advertisements. If you're just starting such a list, ask purchasing agents and contractors from nearby communities (including cities) for their recommendations.

Solicit bids energetically.
- In addition to the advertising required by law (which may have little effect on producing the best bids), send out solicitations to every relevant firm on your resource list.

Not only nationally, but internationally!
- If you're not receiving the volume — or quality — of bids you desire, ring up firms that haven't responded and encourage them to participate.
- A related technique is to gather informal quotes on biddable items to determine the present market. Encourage those with the most competitive prices and dependable service to submit a bid.
- Employ back-up strategies, such as putting the word out among contractors or other associates that you're seeking bids.
- Hold a pre-bid informational meeting and invite prospects to the site. The more contractors know about your project, the more realistic their bids will be.
- If possible, do **not** require short completion times or penalties. A longer time frame will result in lower bids.

Where it's cost-effective, sacrifice convenience or offer the town's own services to make the deal more attractive to bidders.
- Arrange to take delivery in slow times of the year (such as winter in the north) in order to get a manufacturer's lowest prices.
- Can the town arrange transportation from a major rail hub or trucking center to shave some money from a shipper's cost? This needs to be looked at carefully so what you make on the beer you're not losing on the peanuts. But if, for example, you can connect with a local trucker who will be passing that way "deadhead" (without cargo), you might convince him to pick up your shipment and save.

Most statutes stipulate that the award must go to the "lowest responsible bidder." "Responsible" can be a fairly elastic term, but among lawyers the rule of thumb is this: What's reasonable and rational? (One reason for **rejecting** a bid may be that it's too **low**

— not rational or responsible!)

The word "responsible" is not usually defined by statute. As a general rule, courts have supported those who are likely to do a faithful, conscientious job and are financially viable. Elements of responsibility cited by courts also include accountability, skill, experience, ability, judgment, moral character (including prior criminal activity), integrity, and track record on previous similar contracts. Generally, the courts will not substitute their judgment if the municipality's decision was made on a rational basis.

It's essential, of course, that the municipality have documentation **before** reaching the decision to award. It won't help the case if the evidence was assembled after the fact. And just because "everybody knows" and the board "has good reason," there is no substitute for a **written** record.

The inclination to litigate seems to vary directly with the dollar amount of the contract. The more money at stake, the greater the likelihood that rejected bidders will try to get a piece of it. And the threat of a lawsuit is sometimes used as a scare tactic to convince the municipality it can't afford **not** to award to that firm. Such a threat is also more likely if the municipality is on the verge of an extensive building program.

Because so many lawsuits arise over this issue, it is wise to make sure that a rejected low bidder be given an opportunity to present his/her case. Municipalities **can** win even though they bear the burden of proof, but they should be aware that the legal costs can be relatively high. In the long run, accepting the lowest bid seems the most **practical** thing to do. (Be careful whom you choose, however. The local government is also subject to litigation for any project delay which occurs, regardless of whether the delay was in fact caused by one of the prime contractors. The cost of such lawsuits can be exorbitant.)

Remember that the municipality must always reserve the right to reject all bids and then bid again.

COOPERATE WITH OTHER GOVERNMENTS

A New York State booklet on government options begins, "Deficits, federal retreat, and consensus against new taxes collide with continuing constituent expectations for maintenance of government services." Indeed. More simply, local governments find it increasingly difficult to do what needs to be done. Achieving long-term viability is a challenge for every system, but small systems face additional obstacles. They lack economies of scale, they have to spread their costs over a relatively low number of ratepayers, and they frequently have difficulty attracting and keeping highly-qualified staff.

The word is spreading slowly that there **are** a variety of other options for a locality: negotiating intergovernmental agreements, contracting for services with private firms, consolidating management, transferring ownership, physically merging systems, and raising user fees and/or assessments.

While all possibilities should be investigated, it is the first one, partnerships, that may hold the most promise for very small towns. This is an arrangement within the control of the entities involved, and perhaps more readily arrived at because the negotiation may be done on a more expeditious and friendly basis.

By sharing goods, labor, equipment and/or services with another municipality, all parties benefit from greatly reduced costs. In most cases, the partners also achieve corresponding improvement in performance. Partnerships can provide increased convenience: one town may more easily perform a task than its neighboring municipality can.

A partnership may be the answer if there are surplus facilities in one area and deficiency in another, and a partnership may also be appropriate when several parties are attempting to do the same thing — with marginal success. Perhaps the strongest emotional argument for partnerships is that all parties enter the arrangement voluntarily, negotiating in their own self-interest, and retaining a voice in the ongoing implementation. A partnership is an effective way to reconcile the values of individual identity and self-reliance with today's economic realities.

For many years one of the most common voluntary arrangements has been the joint purchasing of salt for northern wintertime highway maintenance in order to achieve quantity discounts. More recently, a number of communities have chosen to share specialized equipment. (A contractor's rule of thumb is that if a big-ticket item is not used for **200** days a year, it's not cost-effective for a municipality to own it!) Others cooperate on laboratory testing, meter readings and billing, other bookkeeping services, etc.

While most communities could find a considerable number of areas in which a partnership would be advantageous — and they may be tempted to do so in their enthusiasm for this strategy — we recommend that the first experience with partnerships be narrowly defined. This is one area where the age-old advice to "crawl before you walk" should be heeded. There are far more items for consideration and negotiation than may be apparent immediately. Course corrections are much more readily accomplished if the scale of the initiative is small. And make no mistake — as participants gain awareness of the implications of joint action, adjustments of the original concept will be required.

A particularly noteworthy application of the partnership approach occurred recently in a rural county in western Maryland, where the state Department of the Environment (MDE) and STEP helped a sanitary district and six small communities to take joint action to improve maintenance of their sewer systems. The Western Maryland Cooperative Utilities Venture (WMCUV) was formed to "make use of opportunities to share equipment, services and resources among participating governmental entities to maximize the efficiency and effectiveness of local operations and preventative maintenance of wastewater systems." WMCUV has received national recognition (including a 1994 Achievement Award from the National Association of Counties) as a model for cooperation among local governments, and undoubtedly holds lessons for other communities. Here we will mention just a few:

The partners identified a common need. They recognized that each system faced similar operation and maintenance challenges, and that individual action was becoming increasingly costly. They concluded that an equipment-sharing arrangement would provide economies of scale in purchasing and allow for a more rational use of community resources. Many types of equipment are needed only occasionally, and with proper scheduling the same device can serve the needs of numerous users.

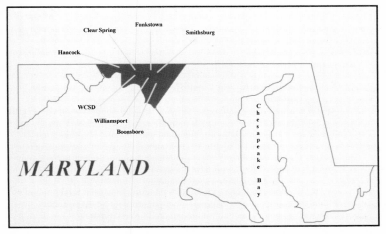

The partners started with a narrow focus. While WMCUV's participatory agreement is flexible enough to include a variety of forms and areas of cooperation, the group felt that, rather than trying to do everything at once, it made sense to begin with a specific, attainable goal in order to solidify the venture. Since each system had problems with inflow and infiltration (I&I), the group decided to purchase specialized equipment that would allow them to address this problem and which none could afford independently.

WMCUV's purpose, policies and procedures are explicitly stated in formal documents. Informal sharing between neighboring systems is fairly common. But as such arrangements get more complex, and especially where significant sums of money are involved, informality becomes a disadvantage. Before agreeing to band together, prospective partners need to know what they're getting into. In forming WMCUV, each entity had the opportunity to air its concerns and shape the result. The group's participatory agreement and statement of policies clarify the rights and responsibilities of the partners and provide for administrative structure, routine maintenance, dispute resolution, rate-setting and other important matters.

A financial package was put together to pay for startup costs. None of the six towns was in a position to contribute substantial sums to get the venture off the ground. Had such a contribution been a requirement for participation, the partnership may not have formed. Fortunately, the towns didn't have to pay anything to join WMCUV. The state contributed $150,000 in seed money, while the county sanitary district provided a $15,000 grant as the local match and an $18,000 loan for additional start-up expenses (the loan will be recouped from user fees).

WMCUV's payment policy and rate structure have been carefully crafted to make the venture financially self-sustaining yet affordable for members. The group decided that each partner would pay a fee, but only when actually borrowing a piece of equipment. The fee has been set to reflect the costs not only of use but of maintenance and amortization as well. This means that money will be available to service the equipment and to replace items as they reach the end of their useful life. In addition, the group has instituted a three-tiered rate structure. Members will pay the lowest fee, since they have put a great deal of time and effort into the partnership. This base fee is about 50% below commercial rental rates. Since the state has invested money in the venture, other entities in Maryland will have the opportunity to use the equipment, but at a higher fee. Finally, entities from other states will be allowed to rent equipment, but at a still higher fee, reflecting the fact that they have made no investment in the venture. In all cases, members will have precedence over outsiders in access to equipment.

The issue of governance was approached with a concern for fairness and equity. WMCUV is governed by an executive board composed of representatives of each of the partners. An elected administrator oversees day-to-day operations. If a member questions a decision made by the administrator, it can appeal to the full board. Four votes are required to carry a measure, and each member has an equal vote, regardless of size. This provision is designed to ensure that no single member will become dominant. This was a concern for some of the towns, each of which has only one system, while the sanitary district serves many communities and thus has a much larger budget.

WMCUV is not simply a partnership between six local entities, but between these entities and the State of Maryland. From the beginning MDE, which regulates wastewater systems, encouraged the venture. MDE reasoned that its investment would pay off handsomely if WMCUV lived up to its promise, since the group included most of the sewer systems in Washington County. In addition to the grant mentioned above, the agency provided direct support through staff engineers who gave advice, helped in the drafting of various documents, and assisted with coordination and logistics.

Appendix B contains a case study and supporting documents that tell the WMCUV story in greater detail. For more information on this partnership, contact WMCUV Administrator Lynn Palmer at 301-223-9416.

Readers who want suggestions on how to facilitate collaborative arrangements between small communities may be interested in a STEP publication titled "Enhancing Small System Viability through Self-Help Partnerships: A Process Guide." Call us at 518-797-3783 for ordering information.

AVOID DUPLICATION

Once again, an investment of time and analysis may save money. Business firms package their services and offer them over and over again, thereby saving on start-up costs for each particular job. This packaging is also done by engineering firms and general contractors. For the community that can afford to turn over most of the responsibility to outsiders, the package contract is probably a good idea: it provides for nearly everything that needs to be done.

But for the locality with little money to spend, such a contract may include needless duplication of services that the community itself can perform. Therefore, carefully read all aspects of the bid or proposal documents and ask for clarification of what is included in each line item. If there's something the local residents can do for less money, the town shouldn't have to pay an outsider to do it.

For example, many construction projects are inspected by the engineering firm. If the locality has a qualified inspector, they could either eliminate or significantly reduce this service by the engineer. Engineering and other contracts frequently contain a line item called "administration." Much of this is usually secretarial work that does not require specialized knowledge. If so, can't this be done by municipal employees?

Does the proposal include a charge for insurance? If so, does the municipality's present policy include adequate coverage? If not, would it be cheaper to get a quote from your own insurance broker? (See pages 149-160 for a more complete discussion of insurance issues.)

The point here is to avoid paying twice for the same thing. Of course, it is always appropriate to question paying even once for

86 / Avoid Duplication

a service that may not be absolutely essential. STEP's rule on this was mentioned earlier, but bears repeating: scrutinize each line item or task, asking, "How can we get equal or better performance for less or no money?"

USE VOLUNTEERS

Since labor can constitute up to 60% of the entire project's budget, reducing that cost can obviously bring huge savings. Those who benefit from the improvement can make an enormous contribution by volunteering their time to work on it. Their efforts not only create or refurbish a needed facility, but they foster community pride and increase local capacity to accomplish even more.

In the mid-1980s The Rensselaerville Institute's work with the Cherokee Nation resulted in the building of a 16-mile rural water line by the Cherokees of Bell, Oklahoma. The savings from their using a volunteer labor force amounted to approximately $300,000! And as described earlier, STEP helped residents of New Jerusalem, Arkansas to extend 4.5 miles of water lines throughout their community. They saved over 55% — and even returned some $1400 in unused grant funds.

Another community that has used volunteers to good effect is Connelly Springs, North Carolina, whose story is summarized on the following page:

Connelly Springs groundbreaking ceremony. Left to right: Dennis Tomlinson, Eric Stockton and Kermit Holshouser.

CASE STUDY:

CONNELLY SPRINGS MAXIMIZES USE OF VOLUNTEERS

Connelly Springs, North Carolina, was developed as a turn-of-the-century resort whose main attraction was the local mineral springs. The community prospered until the mid-1920s when the springs dried up. Ironically, it was water that drew the town back together.

During the 1970s and 1980s futile attempts were made to build a public water system, and in frustration the town re-incorporated in 1989 to explore whether municipal status would facilitate a solution. With such motivation evident, Connelly Springs was chosen as STEP's first demonstration in North Carolina. (Their sparkplug, Mr. Kermit Holshouser, had 4 dry wells on his own property!)

Residents rallied to provide planning, administration, labor and equipment. They worked on 10 Saturdays to lay an average of 500 feet of 6" pipe a day, plus a hydrant every 1000 feet — each hydrant requiring 2 to 3 hours of work. Some weeks they borrowed a backhoe, but mainly they depended on their own farm tractors with loaders or a box sled, hay wagons, and whatever tools they had.

Women and children helped to rake, seed and spread mulch; they also prepared and served lunches and snacks to the workers.

Connelly Springs saved over $67,000 by doing their own construction, plus over $70,000 per year by contracting with the non-profit Icard Township Water Corporation to do the operation and maintenance.

"We're proud of our system," Mr. Holshouser observes. "We're grateful for all the support we've gotten, and I believe self-help is what people need to get back to. I want our project to be a model to show them how."

        Contact: Kermit Holshouser
                        2995 Coldwater Street
                        Connelly Springs, NC 28612
                        (704) 879-9551

Clearly, relying on residents is not possible in every case. A successful volunteer project requires a high degree of attention to organization and supervision, plus special efforts toward motivation and morale. We also caution that a volunteer project may take longer than one done with experienced crews, but for those communities that need to save every penny they can, the delay is an acceptable trade-off.

Here is a list of 34 considerations for using volunteers in a water or wastewater project.

Planning the Volunteer Project

- **Self-help works best when it is prompted by a crisis.** Volunteer commitment is more readily given and sustained when there is a clear sense of urgency. This can sometimes be accomplished by presenting evidence of what **might** befall the community if the improvement is not made quickly. But you cannot cry wolf. Be sure you know what you're talking about.

Kermit Holshouser, Sparkplug of the Connelly Springs water project who later became Mayor to oversee self-help water line extensions to other parts of the town.

- **Citizens need to understand that other funding for execution of the project is not available.** If paid labor were available, few people would choose to volunteer. Residents need to realize that the alternative to volunteer labor is no project at all.

- **Local governments can be just as effective as civic groups in mobilizing volunteers.** Frequently citizens view the role of local government to be larger than that perceived by the elected officials themselves, but residents will respond whether the leadership is from the public or private sector. Local governments must be aware, however, that there is usually a need for different procedures, supervision and even tools when volunteers are involved.

- **Volunteers can be used effectively as managers as well as laborers.** Some 10% to 20% of project costs can be saved by keeping the management local rather that employing an engineering or contracting firm to do it for you. "Outside" businesses must mark up goods and services in order to cover their overhead and profit needs. Even if the local management must be paid it will probably cost less. In sum, if qualified volunteers can be found to manage the project, so much the better.

Management can be divided into two components. **Administration** entails record-keeping and liaison with the outside world, including suppliers, governments, funding institutions and the public. **Construction supervision** involves making sure that the design is being followed; that all equipment, materials and services are delivered satisfactorily and on time; that the work is performed satisfactorily; and that volunteers remain motivated and effective. If necessary, the functions can be split but, if so, frequent and good-natured communication between these two people is **essential.**

- **Get and compare cost estimates for conventional vs. self-help approaches to doing the project.** It is important to know whether self-help will really cost less. If it does, that fact can be used as a strong motivational tool, especially if the total costs are divided by the number of households served to reveal a per-household saving.

- **In establishing costs, be realistic about the difference in performance between volunteer crews and professional ones.** If a professional crew can lay an average of 800 linear feet of pipe per day, figure that amateurs will do about 500. This not only assists in realistic planning, but helps to prevent discouragement among the volunteers if they cannot maintain professional speed.

- **Consider insurance needs and costs.** Workers' Compensation is required to protect all workers, including volunteers. Consult your insurance broker as to whether additional liability coverage is needed. Some communities have asked volunteers to carry their own health insurance. Some propose requesting that volunteers sign a general release from liability, recognizing that they are assuming the risk. We don't recommend the use of waivers, however, for the following reasons:
    ◦ such a release probably wouldn't stand up in court;
    ◦ this undermines the motivation and morale of your work force; and
    ◦ it's irresponsible for the town not to at least cover the liability as minimum recognition of the volunteer's contribution.

    (See pages 149-160 for a more detailed discussion of insurance issues.)

- **Traditional design criteria may need to change.** Different line routings may be necessary to avoid difficult conditions (such as the need to blast through rock, or dig unusually deep trenches) which would be troublesome to volunteers. Needless to say, the design still has to make sense from an engineering perspective. Zigzags don't meet anybody's purpose — except, perhaps, that of the material suppliers. The design can't be so accommodating as to lose effectiveness.

- **Materials may be different, too.** A slightly more expensive pipe that is easy to connect may be cost-effective for volunteers, just as may light-weight materials that can be handled more easily.

- **If volunteers are also participating in project planning, differentiate between the effort and sacrifice required to do physical labor as compared with attending meetings.** While citizen input in planning is desirable and necessary, it does not **substitute** for implementation. Especially if there are to be rewards like public recognition, realize that all work is not equal and some do more than others.

- **If the project is large, break it into manageable pieces.** STEP has found it most helpful to divide a long pipeline into sections that can be installed within a given period. This makes it easier to measure progress, and allows the people who will be served by a particular section to be the ones who work on it.

The Work Process

- **The most effective unit for accomplishing work is the small group.** Compared to a large unit, a small group has more powerful ways of ensuring that its members meet the group's expectations, and provides more effective bonding and cohesion mechanisms. Ideally, those who live near a given section of pipe can and should become responsible for installing it.

- **The key to the group is a crew chief or foreman.** This person is given special training in the construction process and skills required when dealing with volunteers, and then transfers that learning to the crew. If the number of volunteers exceeds a dozen or so, project leaders should rely on a small number of crew chiefs rather than attempt to be responsible for the education of the whole group.

- **It is essential that a knowledgeable, experienced technician be available almost all of the time.** For some groups, a local crew chief may well be sufficient, but for others, a person competent in construction must be brought in. Or the design engineer can be on call. This is to assure the high quality of all the work and to deal with the unexpected where "judgment calls" have to be made. The matter at hand is too important to be treated casually.

This road bore occurred in the Springwood project near Uniontown, PA, one of the few to document the interest of four-legged observers.

- **Some training and skills development are necessary for everyone.** Some concepts can and should be taught in advance, but perhaps the best learning occurs by watching a skillful person perform — even more so if the learner works beside the teacher under field conditions. This process of close study and supervision can not only help to ensure more effective work, it can also provide volunteers with a sense of importance, of capacity for learning, and of contribution to the community.

- **Encourage women to participate at all levels of the project.** STEP is able to document that women are easily the equivalent of men whether in operating equipment, laying pipe or erecting buildings. Project leaders who consider women as suitable only for serving coffee and making telephone calls are depriving themselves of 50% of the community's work force. Women will seldom volunteer for tasks which are beyond their physical strength, but are frequently the first to volunteer to learn a new skill — at least partly due to their lack of embarrassment in admitting they need instruction.

- **The work process must be efficient and fast-paced.** The pace is best set by the crew chief. If he or she works fast, so will everyone else. The opposite is also true.

- **It is also essential that there be quality control.** It is more effective, however, to specify minimum standards rather than ideals. Volunteers will rarely get it exactly right, but they need to know how far off it can be and still be acceptable. It is helpful to designate one worker in the crew as the quality control person: one who checks each joint, trench depth, bedding, etc., and has the authority to stop an individual or crew for necessary corrections. The engineer may also need to make periodic checks in order to certify to the regulatory agency that the project was built according to approved plans.

- **Inventory control is a special problem.** It's not that volunteers are dishonest, but that they can be disorganized. One function which has equal status with physical labor can be that of careful watching and accounting for all inventory storage, flow, and timely replacement when necessary. If a paid person is used for administration, this function might be part of the job description.

- **Safety is an obvious concern.** In addition to a clear and firm announcement of safety rules, it is critical that anyone who violates the rules should be dismissed from the work site immediately. It is the work that dictates the behavior and the precautions — not whether one is paid or unpaid. In a hardhat area, everyone wears one. If anyone arrives smelling of alcohol, he or she must be removed without delay. Once volunteers know the rules and see that they are being enforced, it is unlikely that anyone will disregard safety during the remainder of the project.

- **Expect volunteers to be reliable in attendance.** For the sake of the project, you can't do otherwise. If an expensive piece of equipment and its operator are present while only a part of the crew is there, the work can't go forward. The cost of this might be enormous. Volunteers must understand that they have a continuing obligation to show up when they say they will. If not, like paid staff, they must be replaced.

- **Watch out for too many volunteers!** Too many can be as troublesome as too few! Too many people lead to accidents, a lack of focus, and a general lowering of pace and drive. Treat the scheduling of volunteers as a resource allocation problem: you have a limited number of contributed hours to work with, so make the very best possible use of them.

Morale and Motivation

- **It is generally easier to get people to agree to work on a project before the work begins. The problem is getting volunteers to honor their commitment.** A useful strategy is to make the initial commitment very public. For example, pledges of so many hours for so many weeks can be written on a large sheet at a public ceremony. As the hours are actually completed, that information should be displayed as well. Fewer people will fail to work when they know that everyone's record will be on exhibit. And more people will report for work when they view their original statement as a pledge or promise.

- **It is essential that the work experience be enjoyable.** "Enjoyable" can mean different things to different people. It can mean excitement at having learned a new skill; it can mean a sense of self-worth gained by performing a task well; it can mean social acceptance through cooperative efforts on the job site; and it can mean satisfaction at having participated in an important, historic event.

- **Camaraderie can be built by serving food at the work site.** Lunch, mid-morning and mid-afternoon coffee and snacks let workers know they're appreciated. In addition, refreshments at the site help to keep the crew on the job rather than going home. While good spirit and pleasant company are certainly to be desired at the work site, this must not be allowed to escalate to the point of a party atmosphere. The work is serious business, requiring full attention of everyone present in order to ensure safety.

- **Another morale builder is the presence of a child care service.** Not only are children in a certain amount of danger at the job site, but parents' efficiency is reduced if they are forced to keep one eye on the kids. Establishing a child care service demonstrates the project's encouragement of female workers and gives people who are unable to do manual labor the opportunity to contribute in another way.

- **Worker productivity is an important motivator.** Indeed, a sense of accomplishment is just as important to most volunteers as it is to the managers of the project. If a volunteer can't do a task because something is lacking — knowledge, the right tool or material, the appropriate number of people, etc. — it is reasonable to expect that he or she will become discouraged. If, on the other hand, all these ingredients are in place and the work flows smoothly, volunteers' enthusiasm will become contagious.

- **Allow work crews to be stimulated by friendly competition.** Motivation can soar if small work teams compete to perform similar tasks in a shorter time. The installation of pipes, in particular, lends itself to such rivalry. Carried too far, of course, this can be damaging.

- **Positive reinforcement works much better than negative comment.** Whether workers are paid or volunteer, a little praise and expression of appreciation goes a long way. The praise must be merited and sincere, specifically directed to those who have done a good job at the time they have done it. General appreciation expressed sweepingly to the whole community has little reinforcement value. In fact, it may be **counter**productive! Those who **deserve** the credit and don't get it sometimes become resentful when esteem goes to those who have contributed less — or nothing.

- **Anticipate that the initial enthusiasm will wane as the novelty wears off and the project becomes more routine.** The long middle period of the project requires extra attention to motivational techniques and rewards. Be prepared to unveil special benefits for those who continue to work.

For lasting recognition of Stump Creek, PA's self-help contruction of a water and sewer system, Jane Schautz painted all residents' names on their new storage tank.

- **A variety of special celebrations and perhaps a few tangible mementos or rewards can help a great deal.** For example, we have awarded small, gold-painted, plastic shovels to pipeline workers — but anything else can be effective if it (a) can be displayed, and (b) goes to the real performers. Any presentation should be made as publicly as possible. One way to get a good crowd is to hold the ceremony after a community supper and invite the press and TV cameras. A self-help project makes a great human interest story.

- **Feedback at the project level is essential.** STEP suggests the use of large charts to show progress graphically. Completed sections of a pipeline can be colored in, thermometer-style. You can also keep updated displays comparing actual costs and time with the projected amounts, or a timeline showing tasks and dates of completion can be posted in town hall or a central meeting place.

- **Project leaders must recognize from the beginning that they may well end up doing more than their share of the work.** This is one definition of leadership. Some will have the strength and resolve to continue no matter what the challenges. Others may quit. It's far better to know early who **really** can be relied upon (their past performance is a good indicator). Experienced sparkplugs are not surprised that the job may seem less than glamorous.

- **Voluntarism on one self-help project can lead to self-help in other activities.** For example, while the regulatory agencies prescribe standards for operators of water and wastewater facilities, there is no regulation that says such people have to be paid. Volunteers may well see their own interest being served by continuing to save the community's money once the improvement is made. In addition, the process and skills created to accomplish one public improvement can certainly be transferred to others. Success on subsequent projects is even more likely to occur when residents have already tested their own capacity to work together.

BECOME THE GENERAL CONTRACTOR

It is essential that the community's wishes on priorities and planning be expressed and taken into account. It's also important that the engineering report and design reflect the community's experience and capabilities. But when it comes to construction, the town can also play a decisive and pivotal role **even if its own staff and/or volunteers are not involved as laborers.**

Some towns are too small to commit workers to a self-help project, while residents of other places are either unable or unwilling to become involved. However, it is still possible for the community to provide the management of the project, thus saving the cost a general contractor would have to charge.

Recognize that savings on paperwork and other management functions (environmental review, funding applications, regulatory coordination, etc.) are as real and as important as savings on moving dirt or laying pipe. A good illustration of this point is a small sewer project undertaken by the Village of Smyrna, New York. The mayor personally prepared the environmental review and all the application forms for RDA (now RECD) and SRF funding. She also completed the paperwork for the easements, and even figured out a way to save $1500 in filing fees. This lady doesn't know how to run a backhoe, but the skills she had developed as the manager of a law office resulted in thousands of dollars saved for the community.

Managing a project doesn't mean you do everything yourself. For example, even if the town decides to manage project labor, it can still bring in sub-contractors who will supply it. Frequently very small contractors can't afford to do all the paperwork to bid on government jobs but they are reliable and can often work for less. The municipality's insurance can usually cover them.

Similarly, if the town takes responsibility for materials and equipment, it still has a variety of ways to arrange for purchase (see pages 72-78), and can even borrow or trade with a neighboring community.

The person who undertakes the task of managing the project needs to have most of the qualities of a sparkplug: energy, tenacity, imagination and strong motivation. We'd add one more: experience. While on-the-job training is possible, even inevitable, there isn't enough time to learn all that's needed without jeopardizing the project. It's one thing to understand that some vendors need to be reminded frequently of promised deliveries, but quite another to know that the materials received are the quality and quantity ordered.

Like other self-help techniques, local project administration builds self-reliance that is transferrable to other local undertakings, and provides for in-depth understanding of the substance, challenges and opportunities of local government.

COMMUNICATE!

Obvious? Yes! Frequently overlooked? Also yes! Not only do citizens have an inherent right to know what their elected government is up to, but practical politics requires that taxpayers support the program — especially since it involves large amounts of money.

The Media. Getting press, radio or TV coverage of your activities serves several purposes in addition to providing general information. Media attention also helps with recruitment of additional people as well as reinforcement for the existing workers. To many people, a fact is more credible if it appears in print, and undeniable if it's shown on TV! A self-help project is a natural for media attention since it shows local accomplishment with significant human interest. The media are insatiable in needing new material, so they'll probably be grateful to you for helping them fulfill their own needs.

It's usually a good idea to brief editors and producers early so they can follow your project as it unfolds. Make an appointment to tell them the story and present a packet of materials that document the problem, outline the concepts of self-help, and mention the principal partners so far. In the conversation you should profile your sparkplug — tell why s/he is especially interested and list some of his/her previous accomplishments. Give your own expectations of how the project will proceed.

By all means invite the media to cover your public meetings, but make a real effort to interest them in special events to cover live such as groundbreaking, out-of-town visitors, reaching a particular milestone, etc. Try to notify them at least two weeks in advance. Your press release should give a little of the background, the latest developments to be discussed at the event, name and role of those playing major parts, any new support or contributions and a capsule of interesting anecdotes or quotes.

Don't be reluctant to dramatize the project. Suggest photos of key people engaged in action activities: in a kitchen examining a glass of discolored water taken from the water tap, standing on a porch interviewing a householder during an income survey, in a yard looking at pooled septage. Construction photos are terrific. So are pictures of little kids who are affected or helping out somehow.

Construction photos are appealing in publicity because they communicate action, not just talk.

Show the breadth of participation by including the range of supporters. Not only white male local government officials, but business people and blue-collar workers, and the variety of interested residents including people of different gender, age, education, and income.

Radio interviews also work. Or call up a talk show to relate what's going on.

Other Techniques. In addition to communicating through the media, use direct approaches to keep townspeople informed:

Schedule periodic public meetings. Hold report sessions with ample time for discussion. Encourage questions and different ideas. Be prepared to answer detailed questions while not becoming embroiled in personality conflicts.

Town spirit will be even better if you arrange for people to eat together prior to the meeting. The tiny communities of Edison and Blanchard in northwest Washington State understood that residents needed to hear about project developments, but that it could be done in a friendly, neighborly fashion. They knew that sharing food is such an effective way to develop good feeling that they had a special theme for each of their monthly pot-luck suppers!

Families of New Jerusalem, AR, enjoy a break from their water project. Amy Inman (left, standing on steps) both helped to install pipe and coordinated serving of lunch and snacks to the workers.

Hold neighborhood discussions. It is sometimes both necessary and extremely helpful to arrange for small groups to look at project proposals and impacts. At least one additional person must be present: a group leader who is well informed on concepts and details, one who can answer questions and **transmit**

suggestions to the sparkplug.

Distribute fact sheets. You can also prepare one- or two-page updates periodically that are targeted to your community only. Make them simple in language and format, with news on both the big things and the little ones: for example, total project cost as well as a breezy profile on a volunteer worker. These bulletins must be absolutely accurate — and brief — with lists of facts or short paragraphs of narrative. Copy them on colored paper to make them look special.

For distribution you have the choice of having these flyers:
- mailed,
- delivered house-to-house by kids,
- sent home with kids by the school,
- left in churches, gas stations, bars, laundromats, doctors' offices, etc.,
- posted in conspicuous places,
- passed out at stores, the busiest intersection of town, etc.,
- or circulated by possibly 17 other techniques that you may think up yourselves.

Make speeches to local groups. Most local organizations are desperate for program material, generally being unable to pay more than the price of the speaker's lunch. The members of these groups are likely to be the community activists. Their understanding and approval of the project is essential.

GET THE MOST FROM THE OUTSIDE WORLD

Regardless of how much they can do for themselves, smaller communities will probably need to use some outside resources to carry out local projects. This section of the handbook discusses the best ways to find and use them. Those outside resources may be human (such as lawyers and engineers), or material (such as borrowed equipment), or economic (i.e., loans and grants). But in all cases, some important guidelines apply for connecting with people and institutions outside the community in order to reduce the cost.

As a starting point, we note that when you buy a service or a product — especially advice — you may get more than you think you do. You get a way of looking at the world and a set of approaches to problems that may or may not fit your community. For example, many practices from the outside world which are sold or given to the small town remain urban in nature. Such practices may not fit the values and realities of a small town as well as they fit a city or a suburb.

More specifically, the "conventional wisdom" of the outside world probably doesn't include self-help. The traditional ways of doing water and wastewater improvements rely on contractors, grant packages, and the like. As long as there is money to pay the high costs which result, well and good. But we are now in a new era of fiscal austerity. The reality is that old practices will not necessarily lead to desired results — especially when it comes to making a water or wastewater solution affordable without grant subsidy. Nowhere is the need for a fresh look at the problem more important than in the hiring and use of outside resources.

LEGAL HELP

In many small communities the municipal attorney is the most important person on the local government team. Unlike elected officials, among whom there is a high turnover each year in any given state, the attorney provides continuity, knowledge, and an understanding of what happened previously — and why. In considering self-help, bear in mind that you are not only looking for continuity, however. You are looking for change.

The attorney probably has wider contacts with the outside world, and thus an awareness of resources and events that may be unknown to others. The attorney works full-time, not totally focused on one community to be sure, but constantly engaged in relevant matters — again unlike most of those serving on local boards. For these and other reasons, the municipal attorney's most important product may be timely advice, practicing prevention rather than litigation.

Communities undertaking self-help projects require the active assistance, support and encouragement of their attorney. Above all, they need to be able to rely on someone whose automatic response is not, "You can't do that," but rather someone who will think creatively about legal ways in which they **can** move ahead on their project.

In some cases, it may be necessary to select a new attorney, one who is available nights and weekends when local officials usually meet. And sometimes on short notice. A candidate should also have the following skills or characteristics:

- **Interest and experience in the practice of municipal law.** It is a fact of life that few attorneys start their practice with this knowledge. Unless that was their concentration in law school, they may be doing as well as can be expected if they learn through on-the-job training. With lawyers as with other professionals, ask about their background. Ask what percentage of their practice consists of municipal law, and ask for references from other municipal clients.

- **Aptitude for negotiation and conciliation.** This is very important since so much of what makes self-help projects work involves compromise rather than confrontation.

- **An enabling attitude.** You don't need an attorney who habitually looks for reasons why you can't do something. (That attitude is understandable in that s/he can never be held accountable if no action is taken!) You need an attorney who is prepared to find ways in which you can accomplish your objectives and overcome barriers.

- **Motivation.** Some new STEP communities look for a new, eager attorney who seeks the exposure of being mentioned on a regular basis by the local press. (For that reason it may also help if the attorney is thick-skinned!) It's an added advantage if the candidate is a local resident and has income from other sources since the municipality isn't likely to be able to pay much. The attorney should be fully aware of the legal ramifications of a self-help project, and should be able to suggest other resources. S/he may have some ideas about innovative financing, for example, or know an especially capable resident who might be willing to help. It's obviously a great advantage if the attorney also knows something about other matters, such as accounting, purchasing, or construction.

- **Ability to draft legal documents.** This refers to preparation of local ordinances as well as contracts. Both aspects are necessary.

ENGINEERING HELP

The selection of an engineering firm is to be taken very seriously since the relationship will extend over time and involve significant capital expenditures. With diminishing public projects the engineering profession has grown increasingly competitive; communities can expect to find an engineer who will be pleased to design for local effort and preferences. An engineering firm can help the community save money by explaining the costs and benefits of alternative solutions, and by pointing out ways that use of local resources can be maximized.

From the very beginning, the community should be searching for an engineer who is not only competent and reliable, but willing to work in partnership with his/her clients. Some engineers do this out of "enlightened self-interest." They are aware that if ways aren't found to make local projects affordable, there may not be enough projects to sustain the firm.

However, community leaders must define the scope by clearly establishing what aspects of the project they expect the engineer to complete. This may be facilitated through the use of STEP's Project Task Matrix, available from The Rensselaerville Institute for $3.00 postpaid. The document lists project action steps in tabular form and provides spaces to indicate (1) whether a given activity will be performed by the municipality or the engineer, (2) the estimated completion date for the activity, and (3) the estimated cost of the activity. These worksheets can help the community specify what tasks its people can accomplish competently; the remaining tasks can then be assigned to outsiders.

Both the engineer and the community need to be aware that self-help projects can and frequently do differ in actual design from traditional projects. For example, if a community's employees do not have a great deal of experience, it might be cost-effective to change the location of a pipeline to avoid costly and dangerous blasting of rock. Likewise, it might be helpful if the engineer specified materials that could be lifted without machines.

A community needs an engineer who can present evidence of his/her **experience** in self-help. This may not necessarily be the engineer the community has dealt with in the past. If a firm has **not** previously regarded its clients as **partners** rather than **dependents,** it may not be eager to do so in the future.

Since the very heart of self-help is cost saving, competition among engineers may be to the community's advantage. An engineer's previous knowledge of a particular system in no way guarantees — or is even likely to produce — savings. A firm that has grown accustomed to the community's continued contracts may be inclined to hustle **less**! The community has decreasing leverage with such a firm since there are long-standing personal relationships and habits which may be hard to change.

Moving to a different engineering firm may cause a short-term disruption but it may be worthwhile anyway. A change gives the community a chance to establish a working relationship that is based on the following principles:

- **The community is in charge.** The engineer is employed to make proposals, write reports, prepare the design and provide oversight, but the final decision rests with those who pay the bill. The community needs to know what it wants and then convey those ideas to the engineer. In addition, it always retains the responsibility to apply common sense to the engineer's product. If an engineering plan is unclear, it's the community's job to ask enough questions so that the principles are understood. If the design seems overly complicated or expensive, it is up to the community to insist that simplicity and economy be the watchwords.

- **The engineer is not only a technical resource, s/he is also a teacher.** The engineer should be willing and able to instruct local forces on new concepts and techniques. Such an engineer is very much on the community's side in making the project affordable, and, if necessary, is prepared to personally assist workers in the acquisition of new skills. Some communities add this requirement to the contract.

- **The scope of work to be performed should be fully detailed.** Once the services to be provided are agreed to through negotiations, insist on a "lump sum" or total cost. This should be an "upset" or maximum fee which cannot be exceeded. Additional services not in the scope of work should be provided at an agreed hourly rate, but no additional services should be undertaken without the prior approval of both parties.

One way to achieve both fairness and economy is to negotiate a formula by which the engineer receives a specified bonus or designated percentage of cost **savings** to the project. This is the opposite of one traditional practice we strongly suggest you **avoid:** paying the engineer a percentage of the project's total price. Clearly, this provides no incentive whatever for lowering costs!

While there are a number of checklists available as to what to consider in the actual selection of an engineer, the following cautions should be observed for **any** project:

- Be prepared to spend whatever time it may take to find the right engineer. The process may take longer than you might like, but the wrong choice may be far more painful.

- Insist on meeting with the project engineer, the person who will actually be working with you, not just those who are most skilled in selling the firm.

Issuance of a Request for Proposals (RFP) is the usual way to solicit an engineer. This is a formal announcement of your search, and is sent to a number of firms in your area. Your state regulatory agency can give you names and addresses of those that concentrate on smaller communities with similar problems.

Appendix E presents an outstanding example of an RFP prepared in true self-help spirit by Steve Siddon, sewage treatment plant superintendent in Massena, NY. His very thorough itemization of the proposal's elements includes straightforward language on general information (11 pages), description of professional services (7 pages), information required from proposers (4 pages) and extensive appendices (not included in appendix E).

Where appropriate, Mr. Siddon specifies ways in which self-help shall be involved:

- "The selected engineer must include the operational staff's comments and suggestions as an essential part of the design process." [Page E-256]
- "The Village may opt to utilize its own manpower for portions of the project. The Village may choose this option if opportunities exist to save significant dollars and if the Village has adequate manpower and expertise to accomplish the self-help tasks." [Page E-263]
- "Services that proposers should incorporate in all proposals...include:
    ◦ attending meetings, as necessary with the Village Board, sewer committee and the public. [Page E-264]
    ◦ identifying self-help tasks that can be accomplished utilizing local forces. [Page E-265]
    ◦ providing...preliminary design documents [including] self-help analysis." [Page E-266]

The RFP should mention existing information that proposers may find helpful, and state when and where it will be available for review (you may want to require proposers to make appointments). Such material should include any previous plans, maps, engineering reports, correspondence with regulatory agencies, etc. Mr. Siddon's RFP also states, "It will be assumed that all proposals have taken cognizance of and reflect the relevant data and information contained in such materials" (page E-255), adding that all costs of copying shall be borne by the proposer.

Even though you are requesting proposals for just preliminary engineering, we recommend that you solicit estimates for final design and construction services at the same time for comprehensive comparisons. While it is usually preferable to work with the same engineer through all phases of the project, it is well to execute separate contracts for each phase to give the community flexibility.

The RFP should also announce a preproposal schedule. List the date, place and time of the informational meeting(s), the deadline for written questions, proposal due date, etc.

If this procedure results in a large number of proposals, the normal solution is to eliminate all but three or four of the strongest of them. These proposers are then invited to make an oral presentation to the town council or the sewer committee, with the same questions asked of each applicant. Some of those questions might involve creativity, payment and working relationships, such as:

- What are your ideas on how we can save money on this project?
- How many meetings would you regard as normal for a project like this?
- How often would you expect to visit the site?
- Do you charge for telephone calls? If so, how are they calcu-

lated?
- Do you have another client in this general vicinity who could share your travel costs?
- What are your in-house procedures for quality control of the design?

Do your homework and check out the finalists' references. Ask for the names of the last 10 clients, not just a selected few where things went relatively well. Standard questions include:

- Were there problems with the management, budget, or design of the previous project?
- Were there problems with construction?
- What was the number, percentage, and type of change orders?
- Does the facility operate properly now?

If any of the references involved a self-help project, be sure to ask these additional questions:

- Did the engineer **suggest** ways to save money by using local resources? If so, what were they?
- How much was saved over conventional approaches?
- Did the engineer respond when there was a problem?
- Was s/he available for phone calls?
- Did s/he combine trips to the site with trips to other clients so the travel expenses could be shared?

The final decision can be made less difficult if the RFP indicates the format in which proposers should present their response (see Appendix E, pp. E-270 to E-273). It is also helpful if the reviewers can record basic information on a chart or matrix so that interpretation of proposals is both fair and clear.

Critical elements of comparison include:

- Applicable experience
- Scope of services (work plan)
- Relationship with the community
- Availability
- Flexibility to work with local staff or volunteers
- Innovation: ideas on how to save money
- Cost

If competing firms will charge significantly different amounts, reexamine the work plan to try to understand their reasoning. Do you really need all the services the most expensive firm is proposing? Is the low bidder including everything that's necessary?

After the engineer is selected, spend adequate time in negotiating a fair and equitable contract. Do not take the easy way out by using the engineer's standard "boiler plate" language, as it may well serve the engineer's interests more than your own. Draft your **own** contract, including as many safeguards for the community as possible, and make this the starting point in your negotiations. Consider clauses on approval of key personnel, termination of the contract for cause (including violation of deadlines), professional liability insurance, guarantees of performance, etc.

We recommend attaching to the contract our Suggested Form for Engineers: Monthly Bill to Clients (page 118). You might want to refine the categories for further clarity. Under final design, consider requesting cost breakouts for surveying, borings, hydrologic tests, district formation, and pre-applications for financing. And under construction, consider listing preparation of bid documents, inspection, testing, etc. This provides a work-in/product-out basis for accounting and allows the community to understand the status of work in progress.

Recognize, however, that federal agencies have their own requirements which must be integrated if you anticipate federal assistance. Appendix F is a Farmers Home Administration document that gives instructions for preliminary engineering reports for both water facilities and sewer systems. At press time, these requirements were still in force (although, as discussed on page 137, FmHA no longer exists as such, and its drinking water and wastewater programs are now administered by Rural Economic and Community Development, a new office created under the 1994 reorganization of the U.S. Department of Agriculture).

One final word on hiring the engineer: make sure your attorney scrutinizes the final contract **before** you sign it.

Suggested Form for Engineers: Monthly Bill to Clients

Workplan Task	Staff Time	Consultant	Supplies	Subcontract	Transportation	TOTAL
Preliminary Engineering						
						Overhead/Profit
						Total
Final Engineering						
						Overhead/Profit
						Total
Construction						
						Overhead/Profit
						Total

PROJECT FINANCING

One of the central ideas of this whole handbook is that by undertaking capital projects sooner rather than later, the community benefits not only because it gets the improvement at an earlier date, but because the cost may be reduced significantly by addressing the problem before it gets any worse. Some of those in funding agencies will tell you flat out that it's irresponsible to advocate waiting!

Before listing a variety of ways to finance a project, let's begin by noting that your approach to finding money must be different from the one you have probably used in the past. In the era when grant money was available in relatively large chunks, the common practice was to hire one consultant to get the money and another to spend it. Often the process was in two steps. First, money and work to complete an extensive feasibility study and, second, money and work to do the project itself.

This system had two primary features that must now be called into question. First, it focused almost exclusively on the revenue side. The concept was, literally, of "building" the budget, then looking for someone to fund it. The second feature was reliance on outside experts. Why should communities become involved when the costs of the consultants could also be covered by the grant funds?

These responsibilities must now be reversed. The emphasis must move from "Where do we get the money?" to "How can we keep the costs low enough to make this project affordable?" The latter question recognizes that the best way to raise money is by needing less of it. Note that it can actually cost up to 40% more to get a grant, what with the voluminous documentation and delays involved with processing the application. Emphasis must move from outside handling of the problem to local involvement. All of the self-help strategies suggested here call for the municipal government and the community to do some of the things that they would otherwise pay an outsider to do for them.

A second change concerns sequence of approach. In the past, governments would first identify all possible grant funds, then search about for ways to meet the local share. Now, you start with your own repayment capacity, generally beginning with user fees. By dealing first with the question of how much people can and will pay on a monthly basis, you then have the basis for negotiating the best loan as well as for determining how much grant support you might need.

Let's take the example of a small village that has to replace a decrepit line in its water distribution system. Assume there are 50 households on the system and each has the ability to pay an additional $10.00 a month. By specifying loan conditions (length of loan and interest rate) you can derive local borrowing ability based on user fees. This village could repay a loan of about $49,300 (principal and interest) assuming a 9% interest rate, a 15-year term, and monthly payments.

The people now know what they can afford, but what will it cost? Suppose the county planning office does some rough calculations for the village (for free) and estimates that the project will cost about $59,500. Once the community knows what the gap is, they can seek additional money (for example, a state grant), or find ways to reduce the retail cost through self-help, or some combination of the two.

This example is simplified, of course, but it illustrates the sort of boot-strapping approach that is at the heart of self-help. By using the table in Appendix C you can easily make this sort of calculation for your own real case. This method works for any conventional loan, whether it be for a community water or wastewater project or an individual home mortgage.

A third change is to be aggressive in scoping out all of the costs. For example, many times the engineer's budget does not include interest for the money borrowed during construction.

Self-help tasks must be scheduled to accommodate workers, who may only be available nights, weekends and holidays. In New Jerusalem, AR, many workers even took vacation time.

A note on raising user fees: nobody likes to pay more, but the fact is that water and wastewater costs are increasing due to the need to replace aging infrastructure and comply with stricter regulations. Local leaders may have to educate the public before they can raise rates. They might point out, for example, that the average U.S. household pays far more as a percentage of income for other utilities, such as natural gas, electricity, telephone service, and cable TV. Water and wastewater services are at least as necessary. Leaders might also consider a public display in order to convey the message in a visual manner. One resourceful community assembled 78 gallon jugs filled with water, an impressive sight, with a sign overhead saying, "You now get 78 gallons of water for one penny! Surely it's worth a whole lot more!"

Sources of Capital

Banks. Do comparison shopping. The cost of money is a very significant part of the total project, so do not accept the first interest rate you hear. Rather, find out what other banks would offer.

Stop for a minute to think of the impacts of a lower interest rate. Assume a community borrows $100,000 for 15 years from a bank and makes a monthly payment to cover principal and interest as in a standard mortgage arrangement. The total difference in monies paid back to the bank for using a 10% interest rate rather than 6% is $41,535! The difference in total interest payments increases dramatically if the money is borrowed for a longer term. On the following page is an example of a community that combined self-help cost reduction and a shorter loan term to achieve substantial savings in interest charges.

On the other hand, lengthening the term reduces the amount of the monthly (or quarterly, annual, etc.) payment on a given initial principal at a given interest rate. Often the limiting factor on a community's ability to borrow is cash flow. User fees may be the only source of money to repay the loan, and residents can only pay so much more per month. In such a case, a longer term makes sense, even though more total interest will be paid.

To see the full significance of this, let's take another look at the example of the village water line posed earlier. At 9% over 15 years, the community can borrow $49,300, which is still over $10,000 short of the retail cost. But suppose the village negotiates a slightly lower rate over a slightly longer term: say, 8% over 20 years. Now residents can afford a $59,780 loan — more than enough! (You can check the math using the table in Appendix C). What at first glance appears to be a small change in the terms of the loan turns out to make all the difference!

CASE STUDY:

DOLGEVILLE SAVES BIG ON INTEREST COSTS

Dolgeville is a community of just under 2500 people located in central New York. For nearly a century its drinking water had been drawn from a nearby upland reservoir. However, when the federal Surface Water Treatment Rule went into effect, state regulators told the Village that it would have to filter the water or find a new source.

Local officials were worried about keeping water rates affordable, since developing an alternate source was not feasible, and most filtration technologies are relatively costly. Their fears were confirmed when they looked at the numbers: total project costs would run about $2 million if the job were done the traditional way (having the engineer perform management as well as design and planning functions, hiring contractors to do all the labor, etc.).

Mayor Philip Dahlia and other Village officials contacted the New York State Self-Help Support System to find out how they might reduce the cost through the effective use of local resources. After discussing options with SHSS staff, Dolgeville decided to hire an engineer to design a slow-sand filtration plant, which will last longer than a mechanical plant and have lower operation and maintenance costs. Moreover, project leaders determined that they could save as much as $1 million in construction costs by using municipal employees to do the work.

Since Dolgeville is financing the project through a bond issue, the fact that it needs to borrow only half as much money will have a dramatic effect on total cost savings. Consider: with the lower amount, the Village can afford to cut the pay-back time from 38 to 20 years; since the initial principal is lower and the term is shorter, total interest payments will be lower. At a 7% interest rate, the total interest paid with the self-help approach turns out to be about $2.8 million less than the amount paid if the "retail" method were used!

Residents are the ultimate winners in this case. True, compliance with the filtration requirement will cause a rate hike, but thanks to the energy and determination of local leaders and the Village work force, the increase will be less than half as much as it would otherwise have been.

For more information contact: Mayor Philip G. Dahlia
Village of Dolgeville
41 North Main Street
Dolgeville, NY 13329
315-429-3112

Don't be limited to local banks. Look widely. City banks may be in a position to offer lower rates, although sometimes the reverse is true.

Consider the overall effect. Get the best deal for the taxpayers. If a city bank gives a better rate but your attorney must go there to finalize the arrangements, compare those short-term expenses of travel, meals and lodging with the long-term savings of possibly tens of thousands of dollars from lower interest charges.

Require concessions from your depository. The fact that your official bank has been profiting from the municipality's money over the years is an excellent negotiating tool for insisting on special consideration. Banking is a highly competitive business. Your bank will make concessions if it knows it has to work harder to keep your business.

When negotiating with a bank, your discussion should include at least two elements. First is a reminder to the bank of the importance of local infrastructure (including its improvement) to maintaining the viability of the whole community. Specifically, the infrastructure makes all the houses on a particular line more valuable — many of which the bank may have already mortgaged! This is a very strong argument and deserves emphasis. The second element is rationality. It isn't very persuasive to throw yourself on the mercy of the bank through an emotional appeal. Rather, show that a certain interest rate is all the municipality can afford given your business plan which presents costs and repayment from user fees and, hopefully, other sources. The more clearly the numbers support your case, the more likely the bank will be to do its share. With Appendix C, you can make your own calculations and thus be prepared for your meeting with the loan officer.

It's also good sense to be ready with a contingency or fall-back plan. The bank may balk at self-help in a project, claiming excessive risk as to quality of work, completion, etc. In that case, offer to sign a contract with a reputable construction firm to do the work if the self-help approach fails! Then resume your conversation with the bank about long-range financing, the construction phase no longer at issue. This means you may have won half the battle, but you will have to get construction money somewhere else.

Bonds and notes are listed under banks because that's where the money typically comes from. However, since the Tax Reform Act of 1986 removed many of the incentives they had for holding such debt themselves, banks usually resell bonds to other investors. These investors (individuals, mutual funds, etc.) buy the bonds based on features of government involvement (such as guarantees) and the tax-exempt nature of the interest income from the investment.

The following debt instruments are among the most common. Note that the specific laws governing the use of these instruments by local governments will vary from state to state.

Bond anticipation notes (BANs) are short-term instruments issued by municipalities that expect to receive long-term financing in the near future. Typically they mature within one year, at which time they can be "rolled over" — in effect, you keep paying interest, but not the principal (yet). Depending on the state, BANs can be rolled over for up to five years. As with long-term debt, the interest rate paid on BANs is set by the market. BANs may be particularly appropriate financing mechanisms in these situations: when total construction costs are not yet known, when the issuer believes that long-term interest rates are falling, and/or when the issuer has long-term financing in place but won't actually receive the money until the project is completed.

Tax anticipation notes (TANs), based on anticipated collection of taxes or assessments, are issued for one year at a time but may be renewed so long as they are retired within 5 years of the date of issue. They may be issued in an amount that is no more than total uncollected taxes and assessments for the year, less the amount of outstanding tax anticipation notes and an amount to offset anticipated collection deficiencies. The money raised must be used for the same purposes for which the taxes/assessments are intended.

Revenue anticipation notes (RANs), are similar to tax anticipation notes. Differences: they can be based on assessments other than real estate taxes: "rents," "rates" or "charges" for sewer or water service (the terminology varies across the country), money received from the state or federal government, and other income — such as that from electric light and power plants or other utilities that are owned and operated by a municipality or district corporation.

Capital notes finance all or part of the cost of any project for which serial or sinking fund bonds may be issued. They may be renewed from time to time but mature in two years, with at least 50% maturing in one year. They must be redeemed out of taxes or assessments of the fiscal year.

Bonds are a traditional source of long-term financing. Two major types are general obligation bonds, backed by the general credit and taxing authority of the issuing agency, and revenue bonds, secured by revenues from the project being financed or from another dedicated source. The market generally dictates that bonds be sold for no more than a term of 20 to 30 years, even though state laws often allow the "period of probable usefulness," which may be as much as 40 years. Reasons for issuing bonds for longer terms include the ability to reduce the payments, or to match the life of the bond issue to that of the facility being built, or simply to meet investor demand. However, shortening the length of the bond issue will reduce the total amount of interest paid.

An important limiting factor is the credit worthiness of the community, not likely to be high for rural municipalities in which revenues are low. In certain circumstances local governments can make use of so-called "credit enhancement" strategies to lower interest costs and expand their access to the market. The most common way to enhance credit is through Municipal Bond Insurance (MBI). Prior to marketing the bonds, an issuer applies for an MBI policy. If the application is approved, the insurance company will charge a one-time premium based on a percentage of the aggregate principal and interest due on the bonds from the delivery date to final maturity date. In effect the insurance guarantees timely payment of interest and principal to the bond holders.

There are two problems with MBI, however. First, the percentage used to calculate the premium varies with the credit rating of the community. For an applicant with a high rating, the premium may be as low as 0.35%, but applicants with a low rating could be required to pay four times as much (1.4%). In practice this means that the cost of credit enhancement can exceed its benefit — in other words, it may be cheaper to do without MBI and pay higher interest on the bonds to be issued. Second, MBI is simply not available to jurisdictions that don't have an investment-grade rating from one of the major rating services such as Standard & Poor's. Michael Curley puts the matter bluntly in his *Handbook of Project Finance for Water and Wastewater Systems* (CRC Press, Boca Raton, FL; 1-800-272-7737):

> Those companies which provide credit enhancement... are willing to "enhance" the credit of a water or wastewater system, but they are not willing to substitute their credit for the system's credit. In other words, they will make a

good credit better; but they will not make a bad credit good.

MBI aside, one of the chief objections to bonds themselves is their expense, not the least of which is attorneys' fees. It is essential, of course, to retain bond counsel of proven competence. Many of the most prestigious firms are located in large cities where fees can be staggering, but in smaller cities there are also many bond counseling firms whose fees may be considerably lower.

Bond Banks have been established in several states to help small communities finance infrastructure projects. A bond bank is simply an institution that pools the offerings of many individual issuers so as to gain access to the national market. The advantages of using bond banks have been described by the Environmental Finance Advisory Board in a 1991 publication titled "Small Community Financing Strategies for Environmental Facilities":

> For a community that is unrated or has a poor credit rating, a bond bank's higher credit rating and security provisions usually provide a lower interest rate on the pooled issue. Interest rates are further reduced because pooling smaller bond issues enables diversification, which reduces the risk of default. Pooled issues reduce fees and other up-front issuance costs since each community pays only its share of these fixed costs. Communities also benefit from other economies of scale associated with the pooled issue, such as lower administrative costs as a result of centralized administration by the bond bank. Bond banks also facilitate marketing of issues. Pooled issues are more attractive to underwriters because, being larger, they are easier to sell in the secondary market.

Foundations. The Foundation Center defines a foundation

as "a non-governmental, non-profit organization with funds and program managed by its own trustees or directors and established to maintain or aid social, educational, charitable, religious or other activities serving the common welfare, primarily through the making of grants." A more colorful definition is "a body of money surrounded by people who want some." Foundations have their own funds and some actually make grants to individuals as well as to nonprofit agencies and local governments.

National, general purpose foundations comprise the larger, well-known ones such as The Ford Foundation. Their grants are usually to national or regional operations, and they generally prefer to support innovative or model approaches to solving large-scale problems. The Ford Foundation has provided STEP with the Self-Help Loan Fund of $1.5 million for short-term loans at affordable interest rates to STEP communities.

Special-interest foundations are limited to particular fields or purposes. The R. W. Johnson Foundation, for example, gives only to primary health-care projects.

Family foundations, by far the largest group, were usually started by a single individual. Often these foundations support small projects in certain geographical areas.

Corporation-sponsored foundations are legally separate from the for-profit company, usually having a local orientation and a perspective of self-interest and public relations. For example, The International Paper Co. Foundation assisted a STEP project in the hamlet of New Jerusalem, Arkansas, by providing a $5,000 grant for hookups for the neediest households. IP has a plant in the nearby town of Camden, and several New Jerusalem residents are past or present employees.

Community foundations are public charities with strong local ties and giving patterns. They're frequently administered by a local bank's trust department and governed by a committee of local people.

There are a number of excellent guides to foundations which you can probably find in your public library. We recommend you start with these:

- *The Foundation Directory.* NY, The Foundation Center. Revised annually. This standard reference covers more than 4,000 American foundations with assets of at least $1 million. It includes profiles of each organization's interests, assets and disbursements. Listings are organized alphabetically, geographically and by subject matter.

- *Guide to U.S. Foundations; Their Trustees, Officers and Donors.* NY, The Foundation Center. Revised annually. This comprehensive listing of over 35,000 foundations used to be called the *National Data Book.* It still has an alphabetic listing of foundations in each state, covering thousands of smaller foundations not found in *The Foundation Directory.* It is especially useful for its comprehensive list of directories of state and local grantmakers. The Foundation Center also publishes separate state directories of company-sponsored and community foundations.

- *Corporate Giving Directory.* Rockville, MD, The Taft Group. Revised annually. The 1993 edition describes 607 of the largest corporate givers, each company profiled donating at least $500,000 every year.

How good are your chances? Each year nearly one million proposals are submitted — and that number is growing. According to one informed source, only about 6% or 7% are funded. But don't be put off. This very high turndown rate includes all those who simply sent off mass mailings — as well as the great number of inadequately prepared proposals sent to inappropriate funders.

In order to determine which foundations might really repay your effort, we suggest that you make a phone contact first. In many cases, your request will not fit what the foundation really likes to fund. Don't rely on written guidelines.

Second, focus on foundations with regional or local interests. Otherwise your chances are probably dismal to nonexistent. However, when you're reasonably confident that you've got a fit, spend at least as much effort in reaching and converting a decision-maker within the foundation as you do in preparing your presentation. One way of locating the person who may serve as your internal advocate is to inquire of your attorney or banker. Quite often you'll find that he or she is on the foundation's Board of Directors!

This is not to say that your written proposal isn't important. It is. It needs to be straight-forward, explicit, well-documented and readable. Also recognize that you'll be judged on your track record: what have you already done that indicates you're a good bet?

Corporations. Many corporations like to be good citizens and support worthy causes. Of course, they also like the tax deductions that may come from such contributions. While they generally do not publish comprehensive annual lists of their donations, there are some reference books that should be consulted. Most notable is the *Taft Corporate Directory; Comprehensive Profiles and Analyses of America's Corporate Foundations and Direct Giving Programs.*

Corporations today are far more sensitive than they used to be not only to public relations, but to real issues of social responsibility and commitment. Large, nearby companies should certainly be approached for capital, although it might be easier for them to provide in-kind assistance (see pages 141-144). For both corporations and foundations, location is a key factor. The best bets are those funding sources that have some specific reason to support development projects in your region.

Sometimes public utilities are willing to provide financial support for projects that conserve energy or demonstrate innovative technologies. A small village in northern New York was able to solve operation problems at its wastewater facility thanks in large part to a grant from the regional electric company. The Heuvelton case study follows on the next page.

Individuals. For the lucky communities that include people of wealth and civic consciousness, it should not be overlooked that individuals can — and sometimes do — provide sizeable amounts of money for public improvements. The approach is crucial: it must be made or prepared by the right person. The appeal might be to the prospective donor's ego by naming some permanent installation in his or her honor.

It would be well to check the Internal Revenue Service's latest restrictions on charitable giving before the pitch is made. Depending on the donor's preference, the money could take the form of either a grant or a low-interest loan. Either way, such funds could be the make-or-break ingredient.

Clubs and Organizations. To the extent that local groups understand the problem and also consist of members who would be benefitted by the improvement, raising money for the project might be both a challenge and fun for groups such as Rotary, Lions, Kiwanis, Business and Professional Women and Scouts. Each locality has a sense of what kind of events or promotions are best supported. Beyond an outright appeal for cash donations,

CASE STUDY:

HEUVELTON USES ENERGY GRANT TO FIX SEWAGE FACILITY

Heuvelton was having problems with the aerated lagoon system it used to treat its wastewater. The inefficient mechanical aerators in the activated sludge-style lagoon were becoming very costly to operate, with monthly electricity bills as high as $11,000. Also, the plant had numerous electrical problems, including shock hazards to the operator. Finally, plant staff were finding it increasingly difficult to comply with effluent discharge permit limits.

Concerned by this situation, village officials met with staff from the state's Department of Environmental Conservation and the Self-Help Support System. Mayor Bill Dashnaw had previously received information about an energy-saving alternative aeration system featuring a unique diffuser concept but, along with other leaders, didn't see how the village could afford to finance a retrofit project. At the suggestion of regional Self-Help Coordinator Gary Gunther, the village investigated the Power Reduction Incentive Program offered by Niagara Mohawk, the regional electric utility. Convinced of the potential for large energy savings, Heuvelton submitted an application to NiMo's grant program, then requested a loan from a local bank to round out project financing.

Having solicited bids for the components of the new aeration system, the village purchased them for $96,000. Instead of hiring an outside contractor, village personnel, led by plant operator Gary White, spent 340 staff hours assembling and installing the components at a cost of $2,950. The only other cost was $660 for the rental of a crane. Total project cost was thus $99,610.

The benefits were noticed almost immediately. Electric bills for the first two months after installation were $8,000 and $7,000, respectively, down from a previous monthly average of $10,750. With further adjustments to achieve optimum efficiency, the energy bill was reduced to approximately $6,300.

Niagara Mohawk's power reduction grant paid for 50% of project costs, or $49,805. The bank loan — for two years at 7% interest — paid for the other half. Note that monthly savings in electrical bills now exceed the amount of monthly payments on the loan, and after the latter is paid off, annual savings will be over $40,000!

For more information contact: Mayor Bill Dashnaw
Gary White, Plant Operator
Village of Heuvelton, WWTP
Heuvelton, NY 13654
315-344-2214
315-344-8896

the rule is: Give them fun for their money. It can take the form of food, like a dinner or bake sale; it can be entertainment, like a sports event or concert; a general merchandise sale, like a your-attic-to-mine auction; an arts and crafts festival; etc. The point is that while the project is the direct beneficiary, the community is also rallying round and strengthening its commitment to self-reliance.

Government grants and loans. Remember grants? The award of a grant used to be so predictable that communities would wait patiently for years until their turn came. No longer. But there are several sources you should check out.

Start by contacting the agency in your state that regulates the type of facility (drinking water or wastewater) you're interested in. Ask them what financial assistance they provide, what the eligibility requirements are, and what the application procedure is. To find the address and phone number, see Appendix A, a Directory of State Water and Wastewater Agencies. Even if they refer you to the county or district office, you'll at least get the correct number to call next. The backup procedure is to call the main number in the state capital.

One of the main sources of funding for wastewater projects is the State Revolving Fund (SRF) program. Under the Clean Water Act (CWA), the federal government provides money to capitalize a loan fund in each state, with the state contributing an amount equivalent to 20% of the federal share. The state lends the money to finance eligible projects, and payments on these loans go back into the fund so that they can be re-lent to other communities (hence the fund's "revolving" character).

SRF-eligible projects include construction and rehabilitation of wastewater treatment plants and collection systems; stormwater control; sludge management; nonpoint source pollution control activities; and estuary management. In addition, some states allow SRF loans to be used to refinance existing municipal debt if that debt was incurred for an eligible project constructed after March 7, 1985.

A state has some latitude in how it runs its SRF. For example, states can provide loan guarantees or subsidies for tax-exempt municipal bonds instead of direct loans. Also, each state has its own application procedures, and may have additional requirements beyond those called for in the CWA.

From the local perspective the most attractive thing about the SRF is the interest rates. These can range as high as market rate, but normally are anywhere from one-half to two-thirds of that figure, depending on the financial situation of the borrower. Some states will even provide interest-free loans in hardship cases. As for the term of SRF loans, the maximum is 20 years.

The Council of State Community Development Agencies (COSCDA) notes that "the SRF program has now become the primary financing mechanism for municipal water pollution compliance with [the CWA]." However, be aware that the program has some limitations and disadvantages. First, it is only for wastewater-related projects, although a drinking water SRF program has been proposed in Congress and may already be in place by the time you read this. Second, getting an SRF loan usually entails a comparatively lengthy planning and design procedure, and the money comes with various "strings" attached, such as (in some cases) the Davis-Bacon requirements discussed on pages 62-63. Third, critics say that many SRFs are biased toward larger cities, which tend to have the resources (staff, expertise, etc.) to submit stronger applications than most small communities can. Some states have recognized this problem and have set aside a portion of their SRFs for small towns (but note that applicants must be incorporated municipalities or wastewater authorities).

Finally, since the demand for loans is greater than the supply, most states have priority criteria by which they rank applicants. Low-ranking projects may have to wait years for funding.

For these reasons, COSCDA advises communities contemplating wastewater projects to "consider a variety of financial approaches. Another federal or state program may be more available or better suited for a specific project. A combination of approaches may also be advisable." To find about the SRF in your state, contact the department responsible for wastewater (see Appendix A). The SRF may actually be run by a different agency, but the wastewater staff can steer you in the right direction.

It's also worthwhile to check with your state's economic development agency. It may have grant or loan money available for projects that will have a significant impact on the local economy, even when the motivating force is not purely economic (e.g., the need to protect the public health and/or comply with environmental regulations). By "packaging" the project correctly, you can often make it eligible for assistance from a wider range of sources.

On the federal level, there are two major funders you should know about. One is an office called Rural Economic and Community Development (RECD) which provides loans and grants to help small municipalities (population less than 10,000) finance the construction or improvement of facilities for providing essential services, including drinking water and wastewater treatment. RECD uses a formula based on median household income (MHI) to determine the interest rate and the exact mix of loan versus grant money that will make up the assistance package. The lower the municipality's MHI in relation to that state's non-metropolitan MHI, the lower the interest rate and the higher the proportion of grant funds. However, even for the neediest communities, grants are limited to 75% of project costs.

Note that the loans and grants currently administered by RECD were briefly handled by an entity called the Rural Development Administration (RDA), and for many years prior to that by the Farmers Home Administration (FmHA). However, in late 1994 Congress approved a major reorganization of the U.S. Department of Agriculture (USDA). RDA and FmHA no longer exist, and the programs that each operated have been shifted to other USDA offices. Water and sewer funding programs are now housed in RECD. Be aware that many people still referred to "FmHA financing" even after the money began coming through RECD, and this habit will probably continue until the public becomes familiar with RECD. In any case, your state water or wastewater agency can give current information on this program as well as the phone number of the appropriate RECD office.

The other major source of federal funds is the Community Development Block Grant (CDBG) program of the U.S. Department of Housing and Urban Development (HUD). While HUD provides the money, the states actually administer the CDBG program (the only exceptions are Hawaii and New York which have opted to let HUD administer the program).

Municipalities with less than 50,000 people and nonmetropolitan counties can apply for support for projects that will address local needs in the areas of housing, economic development, or public facilities (including water and wastewater infrastructure). Some states allow multiple jurisdictions to submit a joint application. The primary beneficiaries must be low- and moderate-income residents (the upper limit is typically 80% of the surrounding area's median household income). Grants are awarded for single activities (the category into which most small system projects fall) and for multifaceted projects that address various needs. The amount of a grant varies by state and purpose, but for small communities (population less than 5,000) the average award is $250,000.

Few projects are paid for with CDBG funds alone; in most cases, RECD and/or SRF loans are involved, and a local match is frequently required.

Be aware that competition for CDBG grants is very tough, and in most states applications are accepted only once a year (although some have two or even three rounds a year). Each state has its own application procedures, so interested communities should contact the appropriate state agency. If you don't know which office handles the CDBG program in your state, ask your environmental agency (see Appendix A) or contact the Council of State Community Development Agencies (COSCDA) at 202-393-6435 and ask for their guide called *Financing Water and Wastewater Disposal Systems in Rural Areas*.

Lump-sum payment by users. Your community may include some residents who would be willing to help the project's cash flow by paying their long-term share of improvement costs in a lump sum even before start-up. This does more than help the town; it helps the citizen by eliminating all their interest charges! An attractive inducement for those who can take advantage of it.

If the proposed facilities include significant capacity for future service connections, it is also possible to sell some of that capacity beforehand to industrial, commercial, or residential builders. Those "up front" costs then become part of the development costs (along with other new infrastructure) which are then recovered as part of the cost of the new factory, store or home. This technique has been used very successfully in states such as Florida, where developers often pay a sewer installation fee for each unit to be built.

No matter what other financing is arranged, this option should be pursued. It is worthy of promotion and enthusiastic encouragement by local government officials. Some residents are likely to respond to such an appeal, even though they're aware that the same money might be earning higher interest somewhere else.

Direct legislative appropriation. In many states legislators enjoy the opportunity to nominate projects from their own district for direct inclusion in the state budget. Activities funded in this way are often called "member items." The criteria for worthiness are subject primarily to the legislator's discretion.

Public-Private Partnerships. EPA defines this technique as "a contractual relationship between a locality and a private company that commits both to providing an environmental service." The scope and terms of such arrangements can vary widely, but five basic types have been identified:

- In **contract services**, the public partner owns the facility but hires a private firm to perform some service, such as operation and maintenance (O&M) of a drinking water system.

- A **turnkey facility** is designed, built and operated by a private company that has been hired by a local government. The latter retains ownership of the plant.

- In **developer financing**, private individuals or firms finance an improvement (a sewage treatment plant, for instance) in exchange for the right to build something that will use the new or improved service (e.g., homes, shopping centers, factories, etc.).

- **Privatization** is sometimes used as a synonym for public-private partnership, but in a narrower sense the term refers to cases in which the private sector not only builds and operates a facility, but owns it as well.

- A **merchant facility** is similar to one that is "privatized," but with the additional feature that the decision to build the facility is made by the private partner, not the public entity.

Here are five potential advantages of undertaking a public-private partnership:

- **Access to technology.** Private firms often have greater technical expertise and knowledge of advanced technologies than do local governments. They may also be in a better position than an elected official to undertake the risk of using such technologies.

- **Cost savings** on design, construction and/or operation. For example, a firm that provides similar services to various public partners may enjoy economies of scale in administration, purchasing of supplies, etc. In addition, private companies are often less encumbered by bidding requirements, extensive paperwork and other burdens that add costs for governmental entities. Tax credits used to be a significant source of cost savings but due to tax reforms their importance has declined in recent years. (To learn to what extent a given project may generate tax credits, contact a law or banking firm with proven expertise in public finance and tax-sheltered investment structures.)

- **Access to private capital.** Depending on the type of partnership, the private firm may invest substantial amounts in a facility or service. This is especially true of privatization and merchant facilities. The use of private funds means local governments can reduce or eliminate the need to take on additional debt in order to get the improvement.

- **Delegation of risk.** Such partnerships allow the public entity to shift to the private partner some or most of the risks associated with building a facility or providing a service. To the extent that responsibility for construction delays, performance problems, regulatory violations, and other risks can be transferred to another party, the local government benefits — assuming, of course, that the price charged by the private firm is an acceptable trade-off.

- **Guaranteed cost.** Private firms typically offer to provide a service for a specified period at a set price. Knowing that the cost of a service won't vary for a definite length of time simplifies municipal budgeting and frees up money that a locality might otherwise have to set aside in a contingency or emergency fund.

Before pursuing a public-private partnership, a community must determine whether or not the strategy is appropriate to its situation. One source that can help you decide is a brief article by William Gehr and Michael Brown titled "When Privatization Makes Sense" (*BioCycle*, July 1992, pp. 66-69). For further information on implementation, you may want to consult the EPA document titled *Public-Private Partnerships for Environmental Facilities: A Self-Help Guide for Local Governments* (Publication No. 20M-2003, July 1991). This booklet contains an action checklist for building such a partnership, a discussion of financing, and a list of resources that can provide further information and assistance.

Sources of In-Kind Contributions

Be alert and imaginative as to the existence of other resources that can lessen your need for money. Here are some sources of in-kind contributions you may want to investigate:

Government. In addition to funding, your state water and wastewater agencies may offer technical assistance and useful publications (see Appendix A for addresses and phone numbers). The federal agencies mentioned above, the U.S. Environmental Protection Agency (EPA), as well as Cooperative Extension at your state's land-grant university have a great number of publications that may be helpful.

Closer to home, you should consult nearby municipal governments and water/wastewater operators. The concepts of interlocal cooperation and partnerships were discussed earlier in this book (see pages 79-84), but they are worth mentioning again. STEP has been involved in many projects where one jurisdiction has provided valuable assistance to another. As we have seen, in the New Jerusalem case the county Department of Public Works provided a backhoe to the community so it could install water lines, and in Connelly Springs, a nonprofit water association agreed to operate and maintain the town's water system.

Nonprofit organizations. There are also a number of national nonprofit organizations that provide services ranging from information and referral to onsite technical assistance. These include the American Water Works Association, the Rural Community Assistance Program, the Rural Water Association, and the Water Environment Federation, all of which are described on pages 177-179.

Business and industry. It might be easier for companies to assist the project by providing goods and services rather than money. They may well be able to lend some of their professional staff, such as accountants or engineers, and to offer advice and training. Companies may even be able to lend technicians such as supervisors, purchasing agents and masons. Beyond that, inquire about equipment and supplies — especially if they want to help but don't quite know how.

Colleges and universities. Often you can get assistance with a problem from particular professors who specialize in a relevant field, be it economics, engineering, rural sociology — whatever. A project may be devised by which students undertake information-gathering, analysis, model-building, demonstrations or public education workshops, etc. In the Alfred/Almond, NY, project, a college student ran the income survey! Students receive academic credit for their efforts and the community benefits from their work. Caution: it may take considerable work to define and organize what students actually perform.

Clubs and organizations. Local chapters of the League of Women Voters, Rotary, American Association of University Women, Lions, etc. consist of people who are well acquainted with the fibre and resources of the community. At the very least they should be informed about the project as individuals and groups, but it would be smarter to involve them directly. Delmar, MD/DE acted on that opportunity in a well organized and very successful way. (See page 46 for Delmar's experience.)

Connelly Springs, NC, found a very public way to thank the individuals, groups and government agencies that provided assistance.

Libraries, both public and private. Your local library can secure all sorts of valuable materials through inter-library loan, and may even suggest film, videos, people and other resources. Also inquire about the use of libraries in various governmental departments, colleges, and your engineering firm.

Individuals. If you really put your mind to it, you can probably find ways in which nearly all your residents can participate. It takes time and effort to organize them, but what they provide may make all the difference. Look to individuals for personal service, use of tools and equipment, and guidance in their area of special training.

Self-Help as State Match

Many government financial assistance programs require municipalities to contribute a certain percentage in matching funds (a notable exception is the SRF). An increasing number of states recognize that the value of local work must be taken into account in computing the local match. That work is often crucial to the project's success; without it, the project might not be feasible. The State of Washington's Department of Ecology, for example, has developed a list of items that a town can propose for self-help match in lieu of cash. See Appendix D for Ecology's in-kind policy.

CONTRACTORS

In the discussion so far we have urged local governments to look carefully at their own people and to involve municipal workers to the utmost. We have also emphasized the importance of a realistic appraisal of the actual capacity of local people to do what is needed. Allow and encourage them to aspire to greater accomplishment and increased skills, but not recklessly. The municipality has a responsibility to ensure the safety of its employees as well as to provide a facility that operates with lowest-cost maintenance for a maximum period.

It's most likely that local resources can be applied very effectively in some phases but not others. Your engineer should be able to advise you on what particular skills, experience and equipment are required for each step. Hopefully, the design of the project has already taken self-help into account, calling for simpler technology, less sophisticated (and less expensive) skills, and materials chosen for ease of handling.

But even in self-help projects, some work may have to be contracted out. Engaging a contractor has certain advantages, of course. For example, it allows you to establish a definite cost, set a completion date, provide undivided responsibility, transfer liability from the municipality, and possibly gain access to greater experience.

Finding the right contractor may take considerable effort. The legwork and telephoning alone can take a lot of time. It's customary to begin by asking a large number of people to make recommendations. After you develop a list of prospective firms, ask each for references — and check them carefully. Try to find out how the candidates have handled jobs similar to the one you have in mind. Were their bids responsible? (See pages 73-78 on bidding.) You may also want to check with the Better Business

Bureau to see if any complaints have been lodged against the finalists.

After you have selected a firm, you must obtain a formal, written agreement. Assurances that are not in writing are not binding. Be wary, though, of standard forms prepared by professional associations, as the language in such documents is likely to favor the professional over the client, and probably does not consider any aspects of — or protections needed by — the self-help approach.

It would be extremely difficult to devise a standard self-help contract because of the endless differences among municipalities and projects. However, the agreement you use should address at least these points:

- Names and addresses of buyer and seller (the municipality and the contractor). Make sure the firm you are dealing with is the one named in the contract and that the firm's full name, address, telephone number and name of its official representative are clearly shown.

- A detailed description of the nature and scope of the work, including all work that is subcontracted. Also list standards to be followed.

- Identification of authority: who will be in charge?

- Statement of all warranties, explaining exactly what is covered, and for how long.

- Statement of the contractor's liability and property damage insurance, including, if possible, coverage for subsequent deficiencies.

- Specific starting and completion dates, including penalties for failure to meet these dates.

- Price and terms of payment.

You might also wish to specify that the contractor is required to teach certain skills to municipal employees. In that way local workers can perhaps do more of the job themselves the next time.

As to penalty clauses in the contract, they need to be there "just in case," but recognize that such clauses are seldom effective. The municipality has to prove hours worked, damages, etc., which may be time-consuming and hard to do. The practical answer is to work with your contractor as a partner, each getting benefit from the other. Communicate, especially about delays over which neither has control. As the time for the job approaches, reconfirm the dates originally expected. Don't take an adversarial approach until you've sincerely tried everything else.

If changes or additions to the contract are required, make sure these, too, are in writing and signed by both parties prior to the commencement of work.

Depending on the state and the nature of the project, there may or may not be a requirement that the contractor be bonded. While bonding provides additional security, remember that very small contractors may not be able to afford performance bonds. In that case, the contractor's past performance and reputation may be a reliable indicator of his work.

Let's pause a moment to clarify the differences between bonding and insurance. Insurance is a two-party contract between the company and the insured. Bonds involve three parties: the obligee (contractor or sub-contractor), the bonding company, and the municipality. Performance and payment bonds are not subject to cancellation. Insurance is underwritten with consideration of expected losses, whereas surety bonds, a form of guarantee, are issued on the basis of the contractor's ability to do the job. The bonding company does not expect to pay losses. The bonding company may "subrogate" (pay off a debt for another but then become a creditor for reimbursement) against a contractor to recover any losses it does incur under a bond.

An alternative to bonding is to obtain a letter of credit which some attorneys feel is considerably better. It's not only less expensive, but it's "cleaner." Upon certification that the contractor has defaulted, the bank pays. It's that simple. With bonding, there may be extended and painful negotiations with the issuing company in an attempt to collect. The company may search for excuses as to why it should not pay, and lengthy litigation may result.

An additional consideration: be sure to phase or sequence the work so that municipal employees are not working side-by-side on the same task with the contractor's people. Aside from the obvious problem of pay differences between civil service and prevailing wage rates, mixing crews might allow the contractor to claim "lost time" due to work that fails to meet specifications — or work not complete in time for the contractor's operations. In other words, don't let self-help give even the appearance of delay. Instead, before the contractor starts, have the site prepared for his work, then let him come and do it. The municipality may then inspect the work for acceptability, bearing no liability for delays or other excuses.

INSURANCE

While there has already been extensive discussion about identifying and negotiating with attorneys, engineers, and contractors, the same principles hold for providers of any other needed services. Hire them yourself — directly.

Insurance is one professional service that must be obtained, the cost of which is likely to increase in a self-help project. Coverage already held by municipalities may not be broad enough — or high enough — to anticipate construction activities. However, before itemizing particular kinds of insurance let us turn to some preliminaries.

Regarding suits from the public, generally municipalities are not liable if they are providing a public service. This is called the General Duty Doctrine, the rationale being that a municipality couldn't (or wouldn't) operate if it were liable every time it provided police protection or responded to a fire. Case law has established that this doctrine doesn't apply in areas of water and wastewater! Attorneys call it an "illogical exception" to common law, as municipalities have been given special powers to provide these services (regarded as utilities) of their own volition. If these services are provided negligently, of course, the municipality's liability is not limited.

For such litigation two conditions must be met: (a) determination of negligence and (b) evidence that such negligence was the proximate cause of the damage. (There is also no limit to the number of others who can be named as having liability.) There are pressures to reduce liability, partly as a result of the reluctance of insurance companies to continue writing policies for municipalities.

Insurance is just one of several techniques for managing risk. It is actually the most expensive method, and should therefore be used only as a last resort.

There are basically four techniques in risk management that should be considered:

- **Risk avoidance**. Don't do whatever is causing the risk, or don't allow it to occur. For example, the risk of workers contracting asbestosis can be avoided by not using materials containing asbestos.

- **Risk retention** acknowledges that certain risks occur regularly. Funds can be budgeted to compensate for them when necessary, and it may be cheaper to pay per incident than to carry high-cost insurance.

- **Safety and loss control programs** can make a significant difference. Since the cost of insurance is at least partly dependent on losses sustained, anything that can be done to improve worker safety — or minimize loss (such as with a sprinkler system to reduce fire damage) — should be evaluated.

- **Risk transfer**, ordinarily done by buying insurance. There can also be indemnity clauses in agreements or other guarantees that lessen the need for insurance.

There are a number of general types of insurance product that you should know about. The first is **automobile liability and physical damage insurance.** Undoubtedly you are already aware of automobile liability (claims for bodily injury or property damage), comprehensive (any cause except collision) and collision (collision or overturn). There are also some important optional coverages:

- Uninsured motorist (only applies if the uninsured motorist is liable).

- Specified perils is an "all risk" alternative to comprehensive. You name fire, fire and theft, theft and windstorm, and limited specified perils. This is usually less costly than comprehensive. It is possible to cover automobiles with comprehensive and while providing heavier, more durable vehicles with specified perils.

- Collision self-insurance is frequently a good idea for automobiles that are more than four or five years old, since the premium and deductible combined may exceed the value of the vehicle. It may be wise to continue purchasing collision or specific perils on heavy trucks or other expensive equipment if your cash flow can absorb self-insuring for passenger cars or pickups.

General liability insurance is another important type of coverage. Liability issues arise in several contexts:

- Injury to volunteers or employees participating in construction;

- Property damage to equipment donated and/or used at the construction site;

- Donated materials and equipment being defective and thereby causing personal injury to persons at the site;

- Defective construction resulting in post-construction injuries.

A municipality can be held liable for the acts of its agents, officers and employees. Furthermore, an action for injury to persons or property can be brought against the municipality. Thus, insurance is advisable against these types of potential liability.

The basic comprehensive general liability policy, combined with appropriate endorsements, provides the broadest general liability protection available. It covers not only claims awarded against the insured, but the cost associated with investigating and defending such claims, as well. It also automatically covers liability arising from the premises (the garage and yard), as well as operations (the job site), products and completed operations, actions arising from subcontractors, and liability assumed in certain specified contracts (such as lease of equipment).

Broad Form Comprehensive General Liability is a package endorsement, providing numerous additional coverages and amendments to the basic policy. Your insurance agent should advise as to whether particular additional endorsements or this package would be cheaper.

For excavation, additional coverage should be investigated. The usual liability policy specifically excludes coverage for property damage liability arising from explosion ("x"), underground property damage ("u"), and/or collapse ("c"). Excavation has the potential hazards of all three, while plumbing entails underground property damage risks.

Umbrella liability insurance is designed to cover catastrophic losses. It provides additional limits of liability above those in primary policies, but such coverage is not standardized. Does this mean that if coverage is specified in the basic policy, it will also be included in the umbrella? Not necessarily. Again, read carefully — and ask questions.

During the course of construction there is potential property loss involving the materials and equipment on the job site, as well as any partially completed work. This damage could occur from fire, windstorm, flood, collapse, theft, etc. **Builders risk insurance** is designed to insure virtually all property which has been or is to be incorporated into the project, and it remains in effect until the entire job is completed.

Installation floaters are similar to builders risk policies, but they only cover specific materials and work (usually done by a subcontractor) for only a section of the project. Installation floaters may be required when rather expensive equipment or materials, such as generators, compressors, etc., are involved. For most small communities, even those that are hiring subcontractors, the broad builders risk insurance is sufficient — especially if the subcontractor is a named insured.

Both installation floaters and builders risk policies can be written on a "named peril" basis (mentioning particular risks such as fire, windstorm, vandalism, etc.) or for "all risk" (covering everything that is not specifically excluded). All risk has an added feature involving the burden of proof: when loss occurs, it is the insurance company's responsibility to prove that the loss was not contemplated by the policy in order to deny coverage. On the other hand, when named perils insurance is held, it is up to the insured to prove that the loss fell within one of the covered perils.

Frequently builders risk policies will list as exclusions those coverages which communities would be well advised to carry. Some of these exclusions include faulty workmanship/materials and faulty design, testing flood and earthquake, and transit and offset storage. Again, read carefully and weigh the costs against the benefits.

Sometimes policies include a **co-insurance** clause, the purpose of which is to provide an incentive for the insured to cover all property fully. To achieve this goal, a penalty (which may be substantial) is invoked at the time of loss when appropriate values have not been scheduled in the policy. Co-insurance uses the following formula in determining the amount of a loss to pay:

$$\frac{\text{Amount Insured}}{\text{Amount Required}} \times \text{Amount of Loss} = \text{Amount of Recovery}$$

For example, assume that an installation floater covering an air conditioning unit worth $10,000 contains a 100% co-insurance clause. The municipality insures the unit for $8,000. A fire causes $5,000 damage to the unit, and the municipality would recover only 80%, or $4,000, less the deductible. The arithmetic in this case would look like this:

$$\frac{\$8,000}{\$10,000} \times \$5,000 = \$4,000$$

Some companies will agree to remove the co-insurance clause through use of an "agreed amount endorsement," and this is recommended whenever possible.

Like builders risk and installation floaters, the **equipment floater** can be either on a named peril or an all-risk basis. Many underwriters will write an equipment floater without a co-insurance clause, and this should be done whenever possible.

Some unusual exclusions regarding equipment include:

- Loaned equipment
- Borrowed and rented equipment
- Rental cost reimbursement

Be especially on guard for such exclusions as weight of the load exceeding the equipment's capacity; loss or damage to crane or derrick booms while being operated; loss or damage caused by or resulting from explosion, rupture or bursting of steam boilers, steam pipes, etc. In addition, some policies contain an exclusion for damage to automobiles, trucks or other vehicles.

There are, of course, many other types of coverage, a few of which we shall mention here:

- **Buildings and contents coverage** is usually called property insurance, which local governments usually have. This can be purchased on an "actual cash value" (cost to replace the property minus physical deterioration) or "replacement cost" (which does not deduct depreciation). Remember that insurance valuation is quite different from "book value" (the original purchase cost). Insurable values begin with the actual cost in today's marketplace. The book value doesn't change, whereas the base amount for insurable value increases with inflation.

- **Business interruption insurance** is available to reimburse the actual loss sustained resulting directly from interruption of business — limited to the length of time required to repair, rebuild or replace the property that has been damaged or destroyed.

- **Accounts receivable** — if these records are destroyed or damaged in a fire or similar calamity.

- **Aircraft** — to loss incurred through operation, ownership or use of aircraft.

- **Crime insurance** — especially to cover against theft of municipal assets.

Insurance Suggestions. Regardless of the type of insurance you may need, the following pointers can help you purchase the most cost-effective policy.

- Make sure you have coverage for volunteer workers. Do not accept the word of someone who tells you otherwise. If that person is your present agent, get another agent! Health and accident insurance can be obtained from large national companies that offer policies for sports teams, senior citizens' bus trips, etc. If you can't identify such companies through local contacts, call STEP at 518-797-3783.

  Liability coverage should be easily arranged as an "endorsement" to the municipality's present policy at low or even no cost. As with any policy, read the fine print. Make sure the exclusions don't actually nullify the coverage!

- It is not necessary to use one insurance company for all coverages. Quite often the best coverage can be gotten by having several insurers write policies. But it is advisable to combine certain kinds of insurance with one company when coverage may overlap. For instance, it is better to write general and auto liability with one insurer, as it is with property and boiler and machinery insurance.

- It is preferable to use one agent. Part of the agent's responsibility is to provide balanced coverage after obtaining information from the client. A single agent or firm has a better opportunity to perceive overlaps or gaps.

- Communication is extremely important. Keep your agent or broker up to date on any changes in your operation. A meeting should be held each year at least 60 to 90 days before renewal to provide underwriting data, review loss experience and discuss policy changes. These conversations should be documented in writing to assure understanding and a continuing record.

- Combine insurance policies. Generally, costs can be lowered and coverages broadened by combining as much as possible into one policy. For example, insure all buildings on one property rather than buying separate policies for each.

- Use deductibles to control costs. With most lines of insurance, raising deductibles lowers premiums. The trick is to weigh the premium savings against the additional cost of the deductible amounts — based on frequency of loss. If there is relatively high frequency (10 or more times a year) estimates can be developed by looking at past experience. For losses that occur less often, look at the payback period. For example, if you increase the fire insurance deductible by $1,000 and have a loss, how many years of premium savings will it take to recoup the additional deductible? Either way, also consider your ability to pay for a loss. As a general rule, deductibles seem to make the most economic sense in this order: automobile collision, comprehensive auto coverage, equipment floater, comprehensive property damage liability, and fire insurance.

- Review annual audits by the insurance company. In fact, review them very carefully. Before the auditor leaves with information on actual payrolls, receipts, vehicles, etc. to adjust worker's compensation, general liability and auto premiums, request a photocopy of the handwritten work sheet. When the final typed audit is received, compare it to the handwritten copy. Transposition and math errors happen easily, and they may result in incorrect additional premiums.

- Choose one effective date. It is usually best to have one expiration date for all insurance policies. This helps to coordinate coverage, reduces administrative efforts, and makes it easier to bid insurance competitively.

- Avoid January 1 or the beginning of the fiscal year as the renewal date. This is the busiest time of the year for most agents, brokers and insurers. Consequently, you may well get more attention (and concessions from the company!) if you have chosen a different time. The spring, from March to June, frequently works out well since it also avoids vacation time.

- Obtain and keep loss records. When asked to do so, most reputable insurance companies will provide computer reports showing loss experience. This information should be requested, reviewed in detail at least annually, and kept on file so that a loss record for the last five years can always be developed.

- Require certificates of insurance from contractors and subcontractors. You should also maintain these in a "tickler" or suspense file to make sure they are always current. You may be charged an additional premium if you cannot provide certificates at the time of your worker's compensation audit.

- Competitive bidding frequently improves coverage and lower costs. However, do not bid insurance too frequently — usually not more often than every three to five years. Also, limit the number of agents/brokers or companies to simplify proposal review and allow an organized approach to the market. As a rule of thumb, we suggest that no more than four carefully-selected agents/brokers be involved.

Allocate time to the bidding process; be prepared and willing to give as much information as needed. In general, allow 30 days to prepare specifications and accumulate data, and at least another 30 days to evaluate proposals and place coverage. You should research and have ready at least the following:

- five years of loss information;
- receipts and payrolls for the most current year and forecast for the upcoming year;
- an automobile schedule;
- buildings and contents schedule;
- equipment schedule;
- details on large individual losses;
- updated insurable values for buildings, contents and equipment;
- building areas.

- It is preferable to write new specifications rather than to merely turn over present or past policies. Old information might result in continuing past errors.

- Carefully review proposals, separating coverage from price. The best buy may not be the cheapest quote. **It may be a good idea to get a second opinion from another professional** as to coverage and price. Needless to say, such a consultant should have no relationship to the providing company. In fact, it's better if such a person has no relationship to any providing company. Seek proposals for any insurance consultant as you would for other professional services.

Self-insurance. Insurance has become an acute headache for many communities. Premiums have risen so sharply that paying them has become a real hardship if, indeed, municipalities can get coverage at all.

Municipalities already have the power to insure — which implies the power to self-insure. In effect, this is no more than a deductible: retaining part of the risk. Local governments can set up a reserve fund, or they can allocate money in their annual budget. In the latter procedure, unspent funds would lapse at the end of the year and a new appropriation would be required for continued operation. Obviously, municipalities still need catastrophic insurance to cover amounts beyond ordinary experience.

In states where the law provides for cooperation among municipalities, some communities have banded together to form municipal insurance pools. A number of counties already use this device. Each year a budget is established based on assessed valuation or risk exposure and all the towns in the county are welcome to join.

In many states, localities have come together to form a statewide municipal pool for shared risk coverage. In New York State, it's available through the New York Municipal Insurance Reciprocal at 150 State Street, Albany, NY. Phone: 518-456-5181. In North Carolina, the League of Municipalities offers self-funded insurance programs for property liability (as well as workers' compensation and health) for its members only. Contact the League's Risk Management Services office at P. O. Box 3069, Raleigh, NC, 27602, or call 919-833-1876. In other states, inquire at the Governor's Office or the office of the Secretary of State.

TAKE ESSENTIAL ACTIONS BEFORE CONSTRUCTION

This handbook has taken the general form of a series of checklists: those ideas that must be considered to determine whether and how a self-help project can best be accomplished. While the strategies above may constitute only a partial listing, we hope they serve as a springboard to brainstorming about all available options.

We recommend strongly that each strategy be investigated separately and completely in order to learn its local potential. Some strategies can be pursued simultaneously, while others must await the results of a previous step. Most successful projects utilize many strategies, each contributing to overall cost savings. Not until each option is examined fully can one know the absolute bottom line, a cost figure that cannot possibly be reduced further.

Even though the two-dimensional printed page limits us to presenting items one by one, we must emphasize that interactions are absolutely essential — and a number of activities need to occur at the same time. For example, while an early step in the planning process is the preparation of an engineering report, the community should also become aware of what is required by regulatory agencies, and in what sequence. Let's proceed then to itemize some of the products that must be in hand before construction can begin.

Secure an Engineering Plan

Your community's careful search for an engineer has identified an individual or firm that understands local needs and has experience in proposing and evaluating self-help strategies. After the initial conferences with community leaders to define the process and scope of work, the engineer may or may not touch base before the preliminary plan is ready. But with submission of the preliminary plan, the engineer and community are engaged in an even closer relationship as partners.

The community must remember that its goal, the public improvement, is still a local, community responsibility. The engineer is obligated to handle the technical considerations and prepare the overall plan by which this goal can be met, but it is the critical and open discussion between the engineer and community leaders which will determine the appropriate outcome.

A word of encouragement here to those locals who might feel somewhat hesitant to question the engineer's assumptions or proposals. Common sense is not limited to people with professional degrees, and intelligence is widely distributed throughout the population. Therefore, if the plan is unclear to you, ask questions until you understand it. And if you disagree, or have a better idea, don't be intimidated into remaining silent. The engineer is the community's employee; you're paying the bill!

The engineer's preliminary plan and map of the project area might be entirely acceptable to the community as it is first presented. If not, revisions need to be made until your community is satisfied. (Both the engineer and local leaders should already be aware of what will be acceptable to the regulatory agencies. For example, complying with rules on treatment may be a greater priority to the state regulatory agency than distribution line improvements the community may wish to undertake. Obviously, such differences in priorities must be settled before continuing.) Your community should not authorize the engineer to start project design until full local analysis and evaluation produce real, and not superficial, agreement on the plan.

This is the time to make sure that the engineer has seriously examined all appropriate alternative systems and operations questions (and don't accept "boiler plate" language to give that appearance). Probing questions need to be asked. For example, is this for a pre-designed facility (such as a package wastewater treatment plant)? If so, doesn't that suggest that alternatives for local conditions may not have been carefully explored? And if the engineering costs have already been paid by another client, shouldn't your community get a substantial cost reduction? The engineer's responsibility is to design a system that's customized for your particular needs and conditions.

Each proposed option should include major-category costs so that systematic comparisons can be made. The standard reference books used by both contractors and engineers are the Means Guides, a series published by the R. S. Means Co., Inc. of Kingston, Massachusetts (tel.: 617-585-7880). The most applicable one is *Estimating for the General Contractor* by Paul J. Cook, described as "light on theory, heavy on practical estimating methods and ideas." The book costs $35.95, but perhaps you could borrow it from your engineer or through inter-library loan. The point is that project costs are not determined by magic; local people can do their own homework and arrive at costs independently. Bear in mind that these are estimated average costs — not what's happening in the marketplace right now.

After all the cost reductions from each self-help strategy have been calculated and totaled, then and only then can the final project cost be divided by the number of families to be benefitted, arriving at the cost per household for this capital improvement.

Before you plunge ahead, you may want to give contractors the opportunity to bid on the job to determine whether they can meet or beat the engineer's estimate. If bid prices are still unaffordable, you may wish to contract a portion of the job and leave such things as clearing and grubbing, lawn restoration, building a road to the treatment facility, etc. to local forces. But if the cost is still too high, you know that even more self-help will be required.

Construction supervisor Joe Jones and STEP's Rob Hanna guide installation of pipe in NEEDS Project near Hagerstown, MD.

Investigate and Comply with Regulations

It is absolutely essential to notify the appropriate regulatory agencies of your interest in water or wastewater improvements before you do much else. In some states one agency oversees both wastewater and drinking water matters, while in others, each is regulated by a different agency. See Appendix A for addresses and telephone numbers of state agencies. Be aware that approvals from other levels of government (e.g., regional planning board, county health department, etc.) may also be necessary.

In addition, you may need to establish agreements with a large institution or corporation to be included in your system. This is not only to determine rates and services, but to build in assurances and worst-case compensation if the population of the institution changes drastically or the company leaves.

You should also be familiar with the overall regulatory framework in which your water or wastewater system will operate. Each state has the authority to establish legislation governing the treatment, monitoring, etc. of drinking water and wastewater. However, state rules have to be at least as strict as the federal regulations promulgated in connection with the Safe Drinking Water Act and the Clean Water Act.

Communities contemplating water projects should be aware of the Safe Drinking Water Hotline (1-800-426-4791), which can provide regulatory information and referrals. Another excellent source of information is the National Drinking Water Clearinghouse. As its name suggests, the NDWC is a repository of documentation on issues affecting small water systems. In addition, its technical staff can answer many general questions on regulations as well as technical matters, or refer you to a person or agency that can help. The NDWC's sister organization, the National Small Flows Clearinghouse, provides the same sorts of services on the wastewater side. Both can be reached toll-free by dialing 1-800-624-8301.

Regulations change frequently, or course, so for the most up-to-date information, contact your state water or wastewater agency.

Establish the Right Legal Structure

In most states there are multiple routes toward establishing, owning and operating a water or wastewater system. Municipalities usually have the power to own and operate a system. However, if you are constructing or expanding a system that will cross municipal boundaries, or include an unincorporated area, you may need to establish some sort of special district in order to proceed with the project. The procedures for creating a water or sewer district vary greatly from state to state, and space does not allow us to describe all those differences. The following comments are meant to give you an idea of what's involved.

If your state allows new districts to be created by petition of residents, it's likely that you will need to supply a properly-prepared map, plan and engineering report which includes the maximum cost of the proposed project. State law will dictate who is eligible to sign the petition and how it should be filed.

A public hearing may be required, after which the local board or council passes a resolution either affirming or denying the project in whole or in part. If there is to be public indebtedness for the improvement, further review may be required at the state level. Expect to supply additional information such as:

- aggregate assessed property value of the proposed district;
- total assessed property value of the town;
- average full valuation of property in the town;
- amount of budgetary appropriations;
- maximum amount to be borrowed for the proposed project;
- extent of non-payment of taxes in the area;
- tax rates per $1,000 assessed valuation for property in the proposed district;
- assessed valuation of the average homeowner in the district;
- estimated annual cost to the homeowner for debt service, operation and maintenance;

- a schedule of annual charges to each resident; and
- a projected interest rate on the bond issue.

If your state permits the local governmental entity to establish a new district by resolution, the quantity of documents to be submitted to the state may be abbreviated. Possibly only these will be required:

- legal description of the property;
- maximum amount proposed to be spent on the project;
- the method of financing;
- a statement that the engineer's submissions are on file; and
- the time and place of the public hearing.

If there is any doubt about the process required in your state, contact your regulatory agency early to learn the specifics.

Prepare a Construction Schedule

The successful day-to-day operation of a construction project doesn't just happen. It's the result of a considerable amount of planning and monitoring by someone who both knows and cares. Such a someone is quite literally a sparkplug. If the present staff is so small that a construction manager is not already on hand, or if the prospective project is beyond his or her capability, then another person must be brought on board to take responsibility. You can recruit from a variety of sources, the list including but not limited to the following: a retired engineer, a local contractor, someone recommended by — or possibly present staff of — your engineering firm. Once again, make the selection for construction manager with as much care as was used in choosing the engineering firm.

In addition to identifying the person, it's essential to clarify the process: Who has ultimate responsibility — and how are things handled in an emergency? To whom does the manager turn for answers to questions or solutions of problems? Is it the mayor? The board? The engineer? Or all of the above? If the answer is the engineer, the community may expect some additional charges if this amounts to considerable time. The engineer should be providing oversight as it is essential that the facility be built according to design. (If design changes are made without prior approval, the engineer is released from some liability.)

Fairness dictates that the engineer be compensated for unusual involvement during the construction phase. What constitutes additional billing is a matter that should be discussed with the engineer even before construction begins. If it's up to the manager to get help wherever it's available, the others involved should avoid interference — and after-the-fact back-biting.

The availability of the manager also needs to be made explicit. If not a part of the construction crew and on the site continuously, then the manager should be able to visit the site at least once a day — preferably at varying and unexpected times. S/he also should be present for a daily conference with the crew and, when necessary, outside contractor(s). Periodic conferences with the engineer are essential.

In addition, there should be a policy about the downward flow of responsibility. Who takes over in the manager's absence? And with what after-effects?

The usual arrangement is that the senior person on the crew takes over. (That's senior as to experience, not age!) When this happens, of course, the assistant requires back-up and support. Even if mistakes are made, the second-in-command should be encouraged to use his or her best judgment.

The manager's primary responsibility is the coordination of materials, equipment and personnel. That's not easily done. All three must be present at the right time — no matter who has what excuses. Delays can be not only inconvenient, they may lead to lawsuits over liability. The common-sense approach is to use vendors with reliable service (even though they can't always rely on their suppliers), contractors who are reasonable and can work with some flexibility, and trustworthy employees who can be counted upon for occasional extra effort. Don't abuse the goodwill of any of these people. Earn their respect.

Detailed planning must take place far in advance of construction. Work areas must be prepared: surveyed, staked, cleared, etc. Equipment of the correct type and capacity must be available and in proper working order — with backup and maintenance plans all worked out. Materials should be delivered to the site prior to the arrival of the labor crew. The work in each section should be complete, i.e., excavated, backfilled and restored as quickly as possible to limit the municipality's exposure to vandalism, accidents, complaints and legal claims for damages to the public. In other words, a work schedule must be set up and followed. And that includes covering for contingencies: extra workers and equipment on standby to use if necessary.

The planning process for the whole project is not limited to the provision of materials, equipment and personnel, however. It also involves peripheral matters, such as awareness of utility easements: are any new easements required — and do you have recent maps of utility lines already installed in the project area? Have the utilities been notified to come out and mark their lines? (They ordinarily require at least three business days' notice and should respond within 10 business days.) Have all necessary permits been obtained? And are there any problems with cash flow that need to be reckoned with?

170 / Prepare a Construction Schedule

Once again, you can pay an outside firm significant amounts of money to absorb the headaches of coordination. But you can also save a good portion of those big bucks by managing the project yourself.

Obtain Civil Service Approvals

Most states have civil service statutes that govern the hiring and firing of certain categories of municipal employees. However, requirements vary so much from one state to another that it is difficult to give even general guidelines. Instead, we will illustrate the kinds of issues you may have to address by means of a brief list of considerations that apply in one particular state: New York.

In New York, if the construction plan is based on present employees of the local government, and if their present job description already includes the kinds of tasks they will be doing on the new project, civil service approval is usually not necessary. But if the municipality will be hiring new employees or using present staff in new ways, it may need to obtain civil service clearance before work starts.

All counties and most cities in New York have established their own Civil Service Commissions or Personnel Officers whose duty it is to ensure compliance with the state's Civil Service law. Each jurisdiction adopts Civil Service Rules and Appendices, the latter listing all job titles other than competitive class titles (those requiring an examination). Non-competitive jobs do not require a written test, but rather an evaluation of the minimum qualifications of the applicant in comparison with the job specifications. Even temporary employees must be included in Civil Service review. Before any job is created, a statement of duties must be submitted to the local commission or personnel officer, who then classifies the job. A title is assigned and a job specification is either made up or adopted from the existing categories. The municipality may then create the position: it now has a job slot.

Applicants are then recruited, and when the local government has a nominee, the name and qualifications are submitted to Civil Service. Generally the last two steps are compressed, such that the community applies for both nomination and appointment at the same time.

For many self-help projects, most additional personnel are laborers. Even though this job classification does not have minimum qualifications (you can hire almost anybody), the state's Civil Service procedure must be followed. If not, the name(s) of the new worker(s) will be discovered through submission and Civil Service certification of payroll. If this is the case, work may be interrupted until permission is granted.

New York's Civil Service law allows temporary positions under the following conditions:

- To satisfy terms of a grant.
- For a specific short-term project.
- For an encumbered position. (For example, if a permanent employee is appointed to a higher position and is waiting to take the appropriate Civil Service test, that employee's appointment in the new position is designated as temporary. This is to allow flexibility. If that person should fail the test, he/she is still able to return to the job held previously.)
- For professional services, such as an engineer or construction manager.

In New York, temporary employees can be carried for up to 18 months, depending on the type of position in question. Their reporting, hours, salaries and conditions of employment are prerogatives of the municipality. Temporary employees have no protection for permanent status. Their jobs can be abolished, they can be fired, and they have no right of hearing.

Is all this more red tape? Yes. Does it take a long time? Seldom. With prior notification to the local Civil Service office that the project is under consideration, all the above steps can be done in as little as one day.

In summary, make sure that your checklist of start-up tasks includes contact with your state or local civil service commission. Tell them about your plans (especially any proposed involvement of municipal workers), and ask for guidance on how to comply with any applicable regulations.

PLAN THE TRANSITION FROM CONSTRUCTION TO OPERATION

This section assumes that construction is underway and that procedures are in place to deal with unexpected changes and emergencies. But once again, planning for such eventualities must happen in the beginning. Even though the following items don't all get implemented right away, you've got to ensure a smooth transition from a construction site to an operating facility.

Prepare for Operation and Maintenance (O&M)

There are a number of separate issues that must be addressed:

Rate structure and billing system. Too often the rate schedule is thrust upon the community by the engineer during the design phase — and never changed afterward. In addition to looking for local talent to build the system, the community needs to search for talent to design a billing structure that is fair, progressive, and at the same time encourages water conservation. If rates are graduated, you might consider charging a higher unit price for greater levels of consumption — rather than the traditional structure that rewards waste.

System personnel. Recognize that when you finish building a facility, the operators of your system must be certified (it's really being licensed) by the appropriate regulatory agency. Qualifications in each case are determined by local specifics, such as the capacity and type of treatment of the plant. Spelled out are the requirements for the operator and assistant operator, including basic education, special training course, and type and duration of operating experience. Obviously, for the smaller, less complicated systems, the requirements for operators are easier to satisfy.

Supplies and equipment. Get a checklist from your engineer of what must be on hand — and arrange for timely delivery. Recheck that these items are actually received in good condition, and that an appropriate storage place is ready.

Services. Similarly, a checklist should be prepared for the maintenance schedule. Many states require preparation of a formal and complete O&M manual. For too many small systems, maintenance happens sporadically, if at all. There's no regular flushing of lines, exercising of valves, maintenance of steel tanks, etc. The investment in this kind of prevention is very small in comparison with the cost of a crisis when maintenance is omitted. You will probably need to arrange for any additional services, such as water quality testing. It's not too soon to start comparison shopping.

Office procedures. Do you presently have capacity for added bookkeeping and accounting? How about billing and collection of user fees? If not ready, become so!

Complete All Records and Plans

Efficient operation requires detailed recordkeeping during the entire process, from earliest consideration to final acceptance of the construction. Not only how many people were present at a meeting and a summary of the discussion, but also of its significance — and what happened as a result. And not only a complete set of all invoices and payments (including credits), but a final tabulation of line-item expenditures.

Of particular importance is the keeping of a visual log. Take lots of photographs or video and caption and date them. These are an excellent record for a number of purposes, not the least of which may be litigation.

And probably most important, you must have "as-built" drawings (including date) of every — repeat EVERY — part of the system. Supplementary drawings of installation details will also become increasingly important. Make sure all are clearly labeled and that they are stored in a safe place known to all key personnel.

Record Lessons Learned

Some lessons will not be obvious immediately, but it's very much in your own interest to summarize and evaluate the whole operation and its components as you go along. Hopefully, you've been making course corrections as beginning assumptions proved to be incomplete or incorrect.

The key players as well as a large number of less-involved people can make a contribution to the analysis, for what was learned by one may have been missed by another. Your goal is not only to get different perspectives on what happened and why, but to answer questions concerning the future. For example, what has changed, and what are the impacts of those changes? How can remaining work be improved?

It's important to catch these insights while they're fresh. Try taping large sheets of paper to the office wall, encouraging people to contribute their observations. Filled sheets can be transcribed and kept for final synthesis and judgment.

STEP encourages periodic lessons learned reviews, with at least a half-day scheduled for this at the conclusion of the project.

Celebrate!

The completion of your project certainly deserves a celebration — even if (and maybe especially if) there were some difficult and discouraging moments. You owe yourself the pleasure of self-congratulation! So stage a community dedication around the project's official startup. Plan the festivities, hold an open house. Invite local and state dignitaries to speak. Contact the local media well in advance of the event to solicit their help in publicizing it. (See pages 102-103 for more on involving the media.)

To the extent that you have now successfully completed a significant self-help project, just think how much better the next one will be! You also owe both commendation and appreciation to all those who helped. All the better if the event is shared with the whole community and the media.

It's essential to recognize the symbolic importance of a community's successes. Dedication of a water or sewer project is an opportunity to celebrate and publicly thank all participants.

CONNECT WITH OTHER EXISTING RESOURCES

After the first flush of excitement and pride about your new facility, it's time to savor another satisfaction: you're not alone in dealing with the maintenance of it.

There are networks of people and organizations that can improve your knowledge generally, offer specific guidance and even spare parts, and keep you abreast of new developments in the field. The municipality certainly should sponsor its operators' membership in the state's Rural Water Association (RWA) or the state chapter of the Water Environment Federation (WEF). Each offers invaluable opportunities for training and money savings, and gives operators the chance to meet others who have similar problems. The RWAs work with both small drinking water and wastewater systems, while WEF and its affiliates concentrate on wastewater. For more information contact:

>National Rural Water Association
>P.O. Box 1428
>Duncan, OK 73534
>tel: 405-252-0629

>Water Environment Federation
>601 Wythe Street
>Alexandria, VA 22314
>tel: 703-684-2400

Another organization you should know about is the Rural Community Assistance Program, which works with local officials, community groups and individuals to address problems ranging from water lines to housing needs. RCAP comprises not one agency, but six regional groups united under a national office. Collectively they are often referred to as "the RCAPs," but each one has a distinctive name. For example, the Community Resource Group is the RCAP that serves seven southern states. To find the address and phone number of the RCAP in your state, contact the national office:

> Rural Community Assistance Program
> 602 South King Street, Suite 402
> Leesburg, VA 22075
> tel: 703-771-8636

The American Water Works Association is another good resource. The AWWA is a member organization consisting of individuals, public entities and private firms interested in promoting the quality of drinking water in the U.S. It provides a number of services to members and the public, including training programs, technical advice and education. The AWWA also publishes manuals and a journal. To learn your state's representatives, contact:

> American Water Works Association
> 6666 West Quincy Avenue
> Denver, CO 80235
> tel: 303-794-7711

Previously (page 165) we mentioned two national organizations specifically dedicated to the collection and distribution of information on small water and wastewater systems. The National Drinking Water Clearinghouse and the National Small Flows Clearinghouse, both based at West Virginia University, can be reached at the same toll-free number: 1-800-624-8301. Anyone interested in small systems should, at the least, subscribe to their excellent newsletters: *On Tap* for drinking water, and *Small Flows* for wastewater. In addition, both groups have recently added on-line services that can be very useful to anyone who has access to a computer with a modem. Call the NDWC/NSFC for the latest information on what services are available and how to use them.

The municipality itself should investigate membership in other statewide organizations, such as associations of counties or towns, state conference of mayors, etc. Such groups often provide a variety of beneficial services for their members. On the national level, one of the most respected organizations of this sort is the National Association of Towns and Townships (NATaT), a non-profit membership association that provides technical assistance, educational services and public policy support to local government officials from more than 13,000 communities. Here is NATaT's address and phone number:

>National Association of Towns and Townships
>1522 K Street, NW, Suite 600
>Washington, DC 20005-1202
>tel: 202-737-5200

Communities located in those states that are or have been affiliated with the Small Towns Environment Program have a number of additional resources for technical assistance. At press time that list includes:

ARKANSAS

>David Meador (drinking water and wastewater)
>Water Resources Development Division
>Arkansas Soil & Water Conservation Commission
>101 E. Capitol, Suite 350
>Little Rock, AR 72201
>tel: (501) 682-3978; fax: (501) 682-3991

IDAHO

Bill Jarocki (drinking water and wastewater)
Environmental Research & Analysis Bureau
Division of Environmental Quality
Idaho Department of Health & Welfare
1410 N. Hilton Street
Boise, ID 83706-1290
tel: (208) 334-5879; fax: (204) 334-0417

MARYLAND

George Keller (wastewater)
Engineering & Construction Permits
Water Management Administration
MD Department of the Environment
2500 Broening Highway
Baltimore, MD 21224
tel: (410) 631-3599/3619; fax: (410) 631-3718

Saeid Kasraei (drinking water)
Project Management Division
Public Water Drinking Program
Maryland Department of the Environment
2500 Broening Highway
Baltimore, MD 21224
tel: (410) 631-3714; fax: (410) 633-0456

NEW YORK

Doug Ferguson (drinking water)
NYS Department of Health
Public Water Supply Protection
2 University Place, Room 406
Albany, NY 12203-3399.
tel: (518) 458-6731; fax: (518) 458-6732

Diane Perley (wastewater)
Self-Help Support System
NYS Environmental Facilities Corporation
50 Wolf Road, Room 547
Albany, NY 12205-2603
tel: (518) 457-3833; fax: (518) 457-9200

Ed White (general implementation)
Office of Local Government Services
Department of State
162 Washington Avenue
Albany, NY 12231
tel: (518) 486-4671; fax: (518) 474-4765

NORTH CAROLINA

Eric Stockton (drinking water and wastewater)
Small Community Outreach
Department of Environment, Health & Natural Resources
P.O. Box 29535
Raleigh, NC 27626-0535
tel: (919) 733-6900; fax: (919) 733-9311

SOUTH DAKOTA

Michael Perkovich (drinking water and wastewater)
Office of Facilities Management
Department of Environment & Natural Resources
523 East Capitol Avenue
Pierre, SD 57501
tel: (605) 773-4216; fax: (605) 773-4068

TENNESSEE
Bill Dobbins (wastewater)
Division of Construction Grants & Loans
Department of Environment & Conservation
L & C Tower, 401 Church Street
Nashville, TN 37243-1533
tel: (615) 532-0445; fax: (615) 532-0614

TEXAS
Wayne Wiley (drinking water)
Drinking Water Program
Water Utilities Division
Natural Resource Conservation Commission
P.O. Box 13087
Austin, TX 78711-3087
tel: (512) 239-6059 or 239-6020;
fax: (512) 239-6050

Dr. R.J. Dutton (drinking water and wastewater)
Office of Border Environmental and Consumer Health
Texas Department of Health
1100 West 49th Street
Austin, TX 78756
tel: (512) 458-7111, Ext. 3737

WASHINGTON
Janice Kelley (wastewater)
Water Program
Department of Ecology
P.O. Box 47600
Olympia, WA 98504-7600
tel: (360) 407-6541; fax: (360) 407-6574

Dave Monthie (drinking water)
Financial Needs Assessment
Department of Health
P.O. Box 47822
Olympia, WA 98504-7822
tel: (206) 664-9583; fax: (206) 586-5529

Part Four

The Context

THE NEED FOR IMPROVEMENTS

Authors' note: We are indebted to our EPA partners for the content of this section.

To understand the need for environmental infrastructure improvements, some background on the relevant legal and regulatory framework is necessary. As this revised edition went to press, two major pieces of Federal legislation were before Congress for reauthorization: the Clean Water Act (CWA) and the Safe Drinking Water Act (SDWA).

As part of the reauthorization for the CWA, which was last amended in 1987, the Administration proposes a total of about $13 billion over 10 years to continue capitalization of State Revolving Loan Funds (SRFs). The State contributes a 20 percent match. Loan activities resulting from the Federal appropriations would continue for about 20 years. Through SRFs States make loans to their communities to help pay for municipal wastewater systems and other water protection activities. Loan repayments help replenish the revolving fund.

For the SDWA the Administration proposes a State drinking water SRF with total Federal funding of $4.3 billion spread over 5 years. The SDW-SRF would make loans to communities for projects related to compliance with the SDWA, last amended in 1986.

The CWA is the short name for the Water Pollution Control Act Amendments passed in 1972. The Act aimed to eliminate the discharge of pollutants to U.S. navigable waters by 1985. Its premise was that no one has the right to pollute. The Act set strict standards for treatment and disposal of municipal and industrial wastewater. Where State standards exceeded the minimum national standard, dischargers had to meet the more stringent State standards. And the Act created the multibillion-dollar EPA Construction Grants program to help municipalities build and upgrade wastewater treatment facilities to meet the new standards. Industries were to build treatment facilities with their own resources.

The CWA has been amended several times through the years and though the Act's stringent goals have not been met, it has brought about significant improvements in water quality. Much of this can be attributed to thousands of wastewater treatment facilities built or improved to meet CWA requirements. Between 1972 and late 1994, the public and private sectors invested billions of dollars to build or improve wastewater systems. Of the total, $61.4 billion came from EPA grant money allotted to the States which then distributed the money to local governments, first as construction grants and currently as SRF loans since construction grants funds to the States phased out in September 1990. This money has helped communities build or improve more than 15,000 municipal wastewater facilities.

Despite these vast efforts, water pollution problems persist. Water runoff from urban and rural areas alike as well as growing needs for new construction and upgrades of existing facilities all compound the continuing task of improving water quality nationwide.

Even if Congress approves the Administration's proposed CWA-SRF funding, communities will still need to find other funding sources to meet their wastewater needs. The 1992 EPA Needs Survey estimates $137 billion will be required to fund the projected needs for identified municipal treatment facilities and certain other water pollution control measures that are eligible for SRF funding. The Survey does not estimate all future needs nor those that are not eligible for SRF funding. Nor does it include operation and maintenance costs. The next Needs Survey will be published in 1996.

According to 1992 Census data, the current capital investment in wastewater infrastructure from all members of the infrastructure partnership (Federal, State, and local) is roughly $10 billion a year. This will meet most but not all of the annual needs for construction. Under current financing arrangements, it would appear that to meet all current and future needs all members of the partnership should be investing about $13 billion a year.

For small communities of fewer than 10,000 people, the EPA Needs Survey estimates treatment and collection needs at $13.4 billion with an added $5.4 billion for needs the States identified but that don't meet the EPA documentation criteria. Many small communities lack the resources to adequately document their existing needs. The Association of State and Interstate Water Pollution Control Administrators estimates that almost half of the nation's small communities will need to make substantial capital investments in wastewater systems over the next 10 years. Some of the need for new construction and upgrades is driven by CWA requirements but many municipal plants are aging and would need major improvements anyway. One likely result of increased local responsibility to finance wastewater facilities will be higher user charges. The most recent EPA user fee estimate for the typical household is $150 per year for sewer services. One can only predict that this will go up substantially.

Now let's look at the drinking water side. The intent of the 1974 Safe Drinking Water Act (SDWA) is to protect the public from potentially harmful substances in drinking water. The Act was significantly amended in 1986 and calls for the EPA to set enforceable standards for potential drinking water contaminants. By early 1994 the number of substances regulated was 84 and was expected to increase to 112 by 1995. One of the most important regulations resulting from the amended Act is the Surface Water Treatment Rule, which says that drinking water systems that use surface sources must disinfect and, in some circumstances, filter their raw water.

Here again, the problem is the gap between need and resources. Consider first the challenge for state governments. Under the SDWA, each state is supposed to take responsibility for actually enforcing the standards within its territory — for exercising "primacy". But running a state drinking water program is expensive. There are over 58,000 community water systems (and 134,000 **non**community systems) in the U.S. This means that any given state has to keep track of thousands of systems, making sure that they are in compliance with regulations that are continually increasing in number and complexity. As state legislatures struggle to balance their budgets, drinking water officials find themselves competing for scarce dollars with Medicaid, schools, anti-crime programs, and so on. The result is predictable. In 1993, EPA estimated that the states needed $304 million to fully implement the SDWA program, but that only $142 million was available from state and federal sources. Consequently, most states have not been able to do everything that is called for under the SDWA. A few have even threatened to return primacy to the EPA, which is in no position to run more than a handful of state programs.

Recognizing this shortfall in funds for state drinking water programs, the Government Accounting Office recommended in 1994 that Congress "identify a funding level for the program that (1) will maintain the integrity of the program and (2) better reflects the program's importance in protecting human health." Federal support for state drinking water programs **has** been increasing; it totaled $64 million in 1994, up from $33 million in 1989. But given pressures to control government spending at all levels, it is not at all certain that Congress will continue to increase federal assistance, or that states will be willing and able to come up with the necessary funds on their own.

SDWA implementation is a major challenge for water systems as well as for state regulatory agencies. EPA says that by 1995, public water systems will be paying about $1.4 billion a year just to comply with the standards for the 84 contaminants regulated so far. Other sources put the figure closer to $3 billion. And these regulatory compliance costs are small compared to the capital outlays needed to repair or replace leaky distribution lines, dilapidated storage facilities, worn-out pumps, and other elements of our drinking water infrastructure not directly related to SDWA compliance. The investment required over the next ten to twenty years to address such nonregulatory needs and satisfy increased demand due to population growth is conservatively estimated to be $150 billion.

As with wastewater, it is, of course, the consumer who ends up paying these costs. In a 1993 report to Congress, EPA was clear on this point: "The ability of water systems to pay for SDWA compliance costs will ultimately be based on the ability and willingness of customers to pay the increased water rates needed to finance monitoring and treatment costs." For many households, the burden will be relatively light: a few more dollars per month on their water bill. However, others are already being hit hard, especially those served by small systems, which often lack the economies of scale, access to financing, and technical expertise enjoyed by their larger counterparts. For some customers, compliance will cost several hundred dollars a year.

Many state and local officials see the CWA, and especially the SDWA with its 1986 amendments, as examples of "unfunded federal mandates." Not everyone agrees with this characteriza-

192 / The Need for Improvements

tion — officials in Washington point to the large federal investments discussed above — but Congress, the Clinton Administration and EPA are aware of the need to find ways to ease the burden, especially for small systems. Indeed, a number of reforms were being discussed during the 1994 reauthorization process. For example, the SDWA as currently written requires EPA to develop standards for 25 additional contaminants every three years. Many interested parties, including EPA, would like to see this rigid prescription replaced with a more flexible approach based on scientific assessment of the actual health risks posed by various substances. Also, states may be required to establish source water protection programs so that they can make greater use of the incentive of treatment and testing waivers for communities that institute protective measures. To help finance local compliance, legislators are considering the establishment of drinking water revolving loan funds that states would operate along the lines of their wastewater SRFs.

While some reforms are likely to make it into the reauthorized legislation, small communities shouldn't expect wholesale rollback of regulations. Some of the proposed changes are intended to give communities more options for achieving compliance, as well as a more comprehensive framework for determining what their best option might be. One EPA proposal calls for states to establish programs to assess and enhance the "viability" of drinking water systems. By viability, EPA means a system's financial, technical and managerial capacity to remain in compliance over the long term. The proposed state programs would encourage viability through a combination of enforcement action and technical assistance. They would also place a greater emphasis on "restructuring," which refers to institutional and management changes ranging from the physical merger of systems to cooperation between systems so as to attain economies of scale in one or more areas (e.g., joint purchasing to pool buying power, sharing equipment, use of a circuit-riding operator, etc.). The ultimate weapon in a state viability program would be the authority to order the takeover of a nonviable system by a healthier and probably larger neighbor. (A few states already have statutes that give their regulators this power.)

Small systems are considered particularly vulnerable to viability problems. In addition to the factors mentioned above, the reasons include the low rate base over which to spread the costs of attaining compliance, and the tendency to have part-time operators who may lack the time and/or training to keep up with complex regulations. EPA recognizes the problem, and now runs or funds a number of programs designed to assist small systems. Some of these are discussed elsewhere in this handbook (see pages 50, 165 and 177-179). The point here is that people who live in small communities should understand the concept of viability and how their state plans to ensure it. (Even though the term is used most often in discussions of drinking water, many of the same principles apply to wastewater as well). The discussion of partnerships on pages 79-84 is particularly relevant for systems that may not be viable on their own, and where the community is unable or unwilling to pursue consolidation or annexation.

A good water system is the lifeblood of a community. Whatever can be done to dramatize that fact serves as a building block for the town's viability.

What does it all mean? For the foreseeable future, water and wastewater services are going to get more expensive. It is therefore more important than ever to reduce costs wherever possible. One way to do this is to use self-help methods for community construction projects, for certain operation and management functions, and for partnerships to share costs and services with neighboring municipalities.

We believe that state agencies can also save money by adding a self-help support initiative to their tool kit for achieving compliance. Self-help is not the sort of strategy that can be mandated, but where local residents are willing and able to use it, state regulators often find these advantages over conventional approaches that rely exclusively on grants and/or enforcement action:

- Self-help projects are typically completed in less time due to a reduction in the amount of paperwork and red tape.
- Enabling self-help can be the cheapest way to compliance. It can happen, for example, that the cost to select a grantee is more than the grant itself! As for enforcement, it usually entails time-consuming administrative and legal procedures that do not in themselves produce the necessary improvement to the system.
- While local self-help savings are not captured directly at the state level, it is clear that reduced costs at a municipal level boosts state-level economic vitality.

THE SMALL TOWN AND THE RURAL AREA

In dealing with small, rural communities we need to look squarely at the elements of **definition, scale,** and **values** in such places — and how these may differ from characteristics of cities or suburbs.

The Definition

Here are five problems with the definitions many people now use:

- **Many measuring tools fail to tell us what we need to know.** Even with the U.S. Census, frequently the most authoritative description we have, there is no clear-cut boundary between rural and urban land uses. For example, in distinguishing between "metropolitan" and "nonmetropolitan" counties, the primary criterion used by the Census Bureau is the presence of one or more settlements with more than 50,000 people. However, counties in which a high proportion of the workers commute to a neighboring metro county are themselves considered metro. In many cases, and especially in the Eastern U.S., this definition leads to a metro classification for counties that are dominated by farmland, woods and streams. And a large percentage of U.S. agricultural production comes from such "metropolitan" counties!

The Bureau's second major distinction uses the familiar terms "urban" and "rural," and in some ways it gives a better portrayal of reality, since it is applied to areas smaller than counties. On the other hand, a town with only 2500 people is classified as urban, because a community has to have less than that number to be considered rural. To complicate matters, other federal agencies have different thresholds (RECD, for example, uses 10,000 as the cutoff for "small" communities).

Planners don't always help with the definition, either. They have traditionally used density rates to determine urban, suburban and rural land use. Yet with declining inner-city populations, density there may be reduced so much that it can actually now be **less** than that found in small towns. More affluent suburban developments outside such cities as Denver may require lots of two to five acres, resulting in suburban densities that are also less than those of "rural" settlements.

- **Agriculture is no longer an accurate way to indicate ruralness.** If we look just at agriculture, one of the greatest uses of land in America, we find both that there is now less farming — and that what farming there is has profoundly changed in character. The average farm now represents a larger investment in plant and equipment than is true of the average metropolitan business. Further, while the scale in agriculture continues to increase and consolidate, the scale in manufacturing and other urban business has **decreased** over the past several years. The point is that the presence of a farm is no longer a good indicator that the environment is what we traditionally think of as having rural attributes.

- **There is tremendous variety.** "Rural America" is anything but uniform. When we use the phrase, are we speaking of a wealthy suburb or a migrant camp? The regional variability is enormous. Indeed, the concept of "rural America" is less meaningful than the concept of "urban America," even when limited to one state.

- **Our images may be more myth than reality.** What do we actually find there when we have identified a legitimately rural community? Cities are said to have a fast pace of life, an emphasis on achievement, legal contracts and change. In contrast, the small town is said to be slow-paced, emphasizing characteristics you were born with, interpersonal rather than contractual relationships, and permanence.

While scholars have debated the extent to which these allegations are myth or reality, the truth seems to be that small towns vary greatly in terms of how many differences are still there. Quite often we fail to understand that the differences occur less in the geography of the map than in the geography of the soul.

- **The migration of people away from cities and into rural areas has changed old ways.** Between 1970 and 1980, there was a net migration from metropolitan to nonmetropolitan counties. During the 1980s, this trend reversed, and urban/suburban areas grew much faster than rural ones, some of which experienced an absolute decrease in population. However, the early 1990s have seen the pendulum begin to swing back the other way, and there is an observable trend of migration toward rural areas — especially from suburbs of smaller cities to outlying areas.

Here we will dwell not on what is happening but on what it means. In particular, those moving to small towns are known to be different in certain respects from those who have always lived there. For example, studies indicate that the newcomers tend to have higher incomes, education and job status than those they join. And a sizable number are seeking a good retirement haven.

Also becoming clear are the reasons people are giving for their move. Indeed, they take us to the real meaning of this demographic trend. In brief, factors such as job availability are not as important this time as they have been in any previous American migration. Of increased importance is the desire to escape from urban problems — crime, pollution, overcrowding, high housing costs, and high taxes in particular.

But while migration has generally been welcomed by small towns and rural areas, it has presented four specific problems:

- Some newcomers may seek a level of amenities that can destroy the very quality of life that brought them there in the first place.

- Newcomers tend to want their houses in open spaces rather than in settlements, the places least equipped to provide governmental services and utilities.

- The urban people are bringing city problems with them: there is a rising crime rate in many rural areas.

- Many newcomers may be unwilling to pay for services that they don't need or want. A number of rural areas have seen the rise of "taxpayers' associations," formed by incoming retirees who want to avoid school taxes. These folks may be interested in only the most minimal schools, feeling they have already paid for their own children's education.

The larger challenge, of course, requires the small town to make itself so special that those who came recently can translate their motives from escape (negative) to attraction and nurturing (positive).

Scale and Need

Whatever their population experience, the economic viability of most small towns and rural areas is not pacing their social viability. The loss of agricultural jobs and the movement away from downtown retail stores toward regional shopping malls are two key elements.

Small settlements have no choice but to deal more directly with economic matters. In that quest, three strategies appear especially useful.

- **Import substitution.** Simply put, small towns can wherever possible stop sending their money out of town to buy goods to be imported and do more for themselves at a community scale. Self-help in water and wastewater development falls directly into this category, as do household and community gardens and canneries, home desktop publishing, and other small-scale activities. This emerging pursuit is often called micro-enterprise to suggest the connecting of entrepreneurial acts with life very close to home. In this mode, persons look for very small market opportunities, beginning with what a household or a neighborhood buys.

- **Discovery of small-scale economies.** Much is made of the need for a critical mass and of "economies of scale" that favor large operations. Beginning in the 1980s, however, the "Small is Beautiful" persuasion moved from philosophy to analysis. What it discovered was that whether in manufacturing or in water systems, supposed economies of scale often exist more in theory than in practice and that there are economies that can be achieved at a surprisingly small scale. One reason is the reduction in overhead costs. Another is the avoidance of complexity and the requirement for higher paid persons in a given enterprise.

- **Exploitation of niches.** Some small towns are now acting like enterprising businesses seeking a specific niche within large markets. A few are specializing in such residential areas as extended care and group homes for those with special needs, activities which carry a surprisingly large employment base. Another possibility is to actively seek prototyping opportunities for new technologies in such areas as water and wastewater, where a given set of circumstances and defined boundaries may make ideal conditions.

To summarize, the diversity within rural America is substantial and undeniable. When you have seen one small town, you have **not** seen them all.

Values

The values of rural areas and small towns have played a unique role in America's past. If we believe that history should guide or at least influence our future, then we should understand what that role has been. Starting on a global basis, we should recall that most of the greatest cities in antiquity (Athens, Carthage, and later, Florence) were very small towns by today's standards. What we may consider as provincial or even obsolete because of small size, the Greeks found to be just right for philosophy, commerce and drama.

Closer to home, with the possible exceptions of the church and family, the small town was the basic form of social organization experienced by most people until the early 20th century. Historian Page Smith points out in his outstanding book, *As a City Upon a Hill*, that the small town was more than the common experience of those who wrote our Constitution. The small town may have made the Constitution possible! Says Dr. Smith:

> Without the construction of self-conscious articulate communities that rose far above the organic communal life of

most European towns and villages, the colonists would never have developed the power of common action... In America, the primary covenanted community was expressed concretely in particular small towns; the secondary covenanted community abstractly in the federal constitution.

More recently, a number of studies show that small towns supply leadership to America in numbers well out of proportion to their population. Small town origins appear to make a difference. Consider these thoughts by the late Eric Sevareid, the famous news commentator, after visiting his North Dakota home:

> The memories of a Dakota child are laced with black threads and, for some, the binding is too tight, too painful ever to be unstitched. But the golden threads outlasted the black. Pleasant faces that never die, the creak of saddles and smell of horses, the leafy path to the swimming hole, and mad joy of the circus parade down Main Street. We are all alike, we American men who were boys in the small towns of our country.

From our legendary cities in history to the American constitution to the American leadership contributed by it, the small town has been critically important in American life. If the small town has value to our past, it may be even more important in our future. There are two reasons. First, because it is so numerous. Indeed, it remains the dominant form of the American community. According to the 1990 U.S. Census, there are 3,148 communities in the U.S. with a population of 10,000 or more. By contrast, the communities with **less than** 10,000 people total 20,287 — and nearly half in this latter category are communities sized under 1,000!

This reality is mirrored in infrastructure. EPA data from 1993 indicate that 87% of all community water systems in the U.S. serve fewer than 3,300 persons and that 62% serve fewer than 500 persons. The number of connections (e.g., households) is even lower, of course.

The greater reason concerns not the number of small towns but their distinct value. The small town is a vitally needed counterpart to the erosion and even disappearance of our face-to-face institutions. Many behavioral scientists believe that the loss of ties to place and family and other "primary groups" has made us increasingly dependent on centralized authorities and agencies in all areas of life. This leads to possibilities of totalitarian controls. Social scientist Robert Nisbet has put the matter bluntly in his book called *The Quest for Community:*

> The state can mobilize on behalf of great causes such as war, but as a regular means of meeting human needs for recognition, fellowship, security, and membership, it is inadequate. The only proper alternatives to large-scale mechanical political society are communities small in scale but solid in structure. They and they alone can be the beginning of social reconstruction.

We should note, also, that it is not necessary for the small town ever to have been entirely successful in meeting needs for recognition, fellowship, security, and sense of community for it to have strong value in our present era. The fact that the sustaining symbols of the small town were as much myth as reality does not subtract from their worth. As Lewis Mumford said in 1975 during a talk at The Rensselaerville Institute:

> People who live in villages don't love each other particularly. They are not different in the ordinary affairs of life and they don't differ enormously from the fellows in the cities. They take less stock, perhaps, in what passes for

information in the city but in a sense they are the same people. But in moments of crisis, in birth, in illness or disaster or death, they really display the depth of the human body. They're on hand to help their neighbors. They come to aid without being asked...This is the essential human quality — that beneath all the acquisitions of higher culture, there are certain fundamental things that human beings ought to know about each other. They respect the crises of life and are up to these occasions. That is profound. That is what is important in the small town.

Happily, we can end with a comment that ties value to practicality. The reason that the self-help strategies described in this manual make sense is that they are such a good fit to the small settlement. The capacity for self-reliance runs deeply. If you look hard enough, the chances are you'll find that potential is still in place.

Part Five

Appendices

APPENDIX A

Directory of State Environmental Agencies

Note – In many states, communities with water and/or wastewater problems must seek assistance from regional or county offices. In other states, such assistance is offered by staff based in headquarters in the state capital. Following is a list of state agencies which will assist you or tell you where you sould go for help.

Alabama
Drinking Water
Water Supply Branch of ADM
1751 Congressman W.L.
Dickinson Drive
Montgomery, AL 36130-1463
205–271–7773

Wastewater
Municipal Branch
1751 Congressman W.L.
Dickinson Drive
Montgomery, AL 36130-1463
205–271–7810

Alaska
Drinking Water and Wastewater
Drinking Water/Wastewater Section
Dept. of Environmental Conservation
410 Willoughby Ave., Suite 105
Juneau, AK 99801-1795
907–465–5300

Arizona
Drinking Water
Drinking Water Section
Water Quality Division
Dept. of Environmental Quality
3033 N. Central Ave.
Phoenix, AZ 85012
602–207–2300
800–234–5677 x7995

Wastewater
Aquifer Protection Section
Water Quality Division
Dept. of Environmental Quality
3033 N. Central Ave
Phoenix, AZ 85012
602–207–2300
800–234–5677

Arkansas
Drinking Water
Arkansas Soil and Water
Conservation Commission
101 E. Capitol, Suite 350
Little Rock, AR 72201
501–682–3978

Wastewater
Arkansas Department of
Pollution Control & Ecology
8101 Interstate 30
Little Rock, AR 72219
501–570–2828

California
Drinking Water
Drinking Water Technical
Program Branch
Dept. of Health Services
601 North 7 Street MS92
PO Box 942732
Sacramento, CA 94232–7320
916–323–6111

Wastewater
Regional Wastewater Quality
Control Boards
Water Resources Control
Boards
Box 100
Sacramento, CA 95812-0100
916–657–2399

Colorado
Drinking Water
Drinking Water Program
Water Quality Control Division
Dept. of Health & Environment
4300 Cherry Creek Drive So.
Denver, CO 80222–1530
303–692–3500

Wastewater
Wastewater Program
Water Quality Control Division
Dept. of Health & Environment
4300 Cherry Creek Drive So.
Denver, CO 80222–1530
303–692–3500

Connecticut
Drinking Water
Water Supplies Section
Department of Public Health
150 Washington Street
Hartford, CT 06106
203–566–1253

Wastewater
Department of Environmental
Protection
165 Capital Ave.
Hartford, CT 06106
203–566–5599

Delaware
Drinking Water and Wastewater
Division of Water Resources
Dept. of Natural Resources and Environmental Control
89 Kings Highway
P.O. Box 1401
Dover, DE 19903
Drinking Water: 302–739–4590   **Wastewater**: 302–739–5731

Florida
Drinking Water
Drinking Water Section
Dept. of Environmental Protection
2600 Blair Stone Road
Tallahassee, FL 32399–2400
904–487-1762

Wastewater
Wastewater Facilities Regulation Section
Dept. of Environmental Protection
2600 Blair Stone Road
Tallahassee, FL 32399
904–488–4520

Georgia
Drinking Water
Drinking Water Program
Environmental Protection Division
East Floyd Towers, Suite 1362
205 Butler Street, SE
Atlanta, GA 30334
404–656–5660

Wastewater
Waste Water Program
Environmental Protection Division
East Floyd Towers, Suite 1070
205 Butler Street, SE
Atlanta, GA 30334
404–656–4887

Hawaii
Drinking Water
Safe Drinking Water Branch
Environmental Management Division
Dept. of Health
919 Ala Moana Blvd., Rm.308
Honolulu, HI 96814
808–586–4258

Wastewater
Wastewater Branch
Environmental Management Division
Department of Health
919 Ala Moana Blvd., Rm.309
Honolulu, HI 96814
808–586–4294

Idaho
Drinking Water and Wastewater
Community Programs
Division of Environmental Quality
1410 North Hilton
Boise, ID 83706–1290
208–334–5860

Illinois
Drinking Water
CAS-19
P.O. Box 19279
2200 Churchill Rd
Springfield, IL 62794–9276
217–785–0561

Wastewater
Water Pollution Control
Permits, Planning, and Grants
P.O. Box 19276
2200 Churchill Rd
Springfield, IL 62794–9276
217–782–1654

Indiana
Drinking Water
Drinking Water Branch
Dept. of Environmental Management
100 N. Senate Ave.
P.O. Box 6015
Indianapolis, IN 46206–6015
317–233–4222

Wastewater
Dept. of Environmental Management
Facility Construction Section
100 N. Senate Ave.
P.O. Box 6015
Indianapolis, IN 46206–6015
800–451–6027

Iowa
Drinking Water
Water Supply Dept.
Dept. of Natural Resources
Wallace State Office Bldg., 5th Floor
East 9th and Grand
Des Moines, IA 50319–0034
515–281–5145

Wastewater
Solid Waste Dept.
Environmental Protection Division
Dept. of Natural Resources
Wallace State Office Bldg., 5th Floor
East 9th and Grand
Des Moines, IA 50319–0034
515–281–5145

Kansas
Drinking Water and Wastewater
Bureau of Water
Dept. of Health and the Environment
Forbes Field, Building 283
Topeka, KS 66620
913-296-5500

Kentucky
Drinking Water and Wastewater
Division of Water
Dept. of Environmental Protection
14 Reilly Road
Frankfort, KY 40601
502-564-3410

Louisiana
Drinking Water and Wastewater
Engineering Division
Office of Public Health
Dept. of Health and Hospitals
P.O. Box 60630
New Orleans, LA 70160
Drinking Water: 504-568-5101
Wastewater: 504-568-5100

Maine

Drinking Water
Dept. of Human Service
Bureau of Health Engineering
State House, Station 10
Augusta, ME 04333-0010
207-287-5694

Wastewater
Bureau of Health Engineering
State House, Station 10
Augusta, ME 04333-0010
207-287-5672

Maryland

Drinking Water
Public Drinking Water Program
Water Management
Administration
Dept. of the Environment
2500 Broening Highway
Baltimore, MD 21224
410–631–3702

Wastewater
Engineering and Construction
Program
Water Management
Administration
Dept. of the Environment
2500 Broening Highway
Baltimore, MD 21224
410–631–3599

Massachusetts

Drinking Water
DWS - 9th Floor
Dept. of Environmental
Protection
One Winter Street
Boston, MA 02108
617–292–5770

Wastewater
DWPC – 8th Floor
Dept. of Environmental
Protection
One Winter Street
Boston, MA 02108
617–292–5673

Michigan

Drinking Water
Division of Water Supply
Dept. of Pubic Health
3423 M.L. King
P.O. Box 30195
Lansing, MI 48909
517–335–9216

Wastewater
Surface Water Quality
Division
Dept. of Natural Resources
P.O. Box 30273
Lansing, MI 48933
517–373–1949

Minnesota

Drinking Water
Dept. Environmental Health
P.O. Box 59040
Minneapolis, MN 55459–0040
612–627–5100

Wastewater
Municipal Section
Water Quality Division
Pollution Control Agency
520 Lafayette Rd.
St. Paul, MN 55155
612–296–6300
800–657–3864

Mississippi

Drinking Water
Division of Water Supply
Dept. of Health
P.O. Box 1700
Jackson, MS 39215–1700
601–960–7518

Wastewater
Land and Water Resources
Dept. of Environmental Quality
P.O. Box 10631
Jackson, MS 39204-0631
601–961–5200

Missouri

Drinking Water
Public Drinking Water Program
Dept. of Natural Resources
P.O. Box 176
Jefferson City, MO 65102
315–751–5331

Wastewater
Water Pollution Control Program
Dept. of Natural Resources
P.O. Box 176
Jefferson City, MO 65102
315–751–1300

Montana

Drinking Water and Wastewater
DHES - Water Quality Division
Health and Environmental Sciences Dept.
1400 Broadway
P.O. Box 200901
Helena, MT 59620–0901
406–444–2406

Nebraska

Drinking Water
Division of Drinking Water & Environmental Sanitation
301 Centennial Mall South
Lincoln, NE 68509–5007
401–471–2541

Wastewater
Dept. of Environmental Quality
Wastewater Facilities Section
P.O. Box 98922
Lincoln, NE 68516
402–471–4200

Nevada
Drinking Water
Wells, Dams, Drilling & Water Rights
Division of Water Resources
123 West Nye Lane
Carson City, NV 89710
702–687–4380

Drinking Water Quality
Bureau of Health Protection Services
505 East King Street, Rm 103
Carson City, NV 89710
702–687–4750

Wastewater
Water Pollution
Division of Environmental Protection
Capital Complex
Carson City, NV 89710
702–687–4670

New Hampshire
Drinking Water and Wastewater
Dept. of Environmental Services
Bureau of Wastewater Treatment
6 Hazen Drive
P.O. Box 95
Concord, NH 03302–0095
603–271–3504

New Jersey
Drinking Water
Division of Water Resources
Dept. of Environmental Protection
CN–426
Trenton, NJ 08625
609–292–5550

Wastewater
Division of Water Quality
Dept. of Environmental Protection
401 East State Street
P.O. Box CN–029
Trenton, NJ 08625
609–633–3869

New Mexico
Drinking Water and Wastewater
Environment Dept.
P.O. Box 26110
Santa Fe, NM 87502
505–827–2834

New York
Drinking Water
Department of Health
Division of Environmental
Protection
Bureau of Public Water Supply
2 University Place
Albany, NY 12203-3313
518–458–6731

Wastewater
Self–Help Support System
Environmental Facilities Corp.
50 Wolf Rd., Rm. 547
Albany, NY 12205–2603
518–457–3833

North Carolina
Drinking Water
Division of Environmental
Health
P.O. Box 29536
Raleigh, NC 27626–0536
919–715–3232

Wastewater
Division of Environmental
Management
P.O. Box 29535
Raleigh, NC 27626–0535
919–733–7015

North Dakota
Drinking Water
Division of Municipal
Facilities
Dept. of Health
Box 5520
Bismarck, ND 58502–5520
701–221–5211

Wastewater
Division of Water Quality
Dept. of Health
Box 5520
Bismarck, ND 58502–5520
701–328-5227

Ohio
Drinking Water and Wastewater
Environmental Protection Agency
Division of Drinking and Ground Waters
1800 Watermark Drive
P.O. Box 1049
Columbus, OH 43216–1049
614–644–2752

Oklahoma
Drinking Water and Wastewater
Dept. of Environmental Quality
Water Quality Division
1000 NE 10th Street
Oklahoma City, OK 73117–1212
405–271–5202

Oregon
Drinking Water
Drinking Water Program
Health Division
Dept. of Human Resources
P.O. Box 14450
Portland, OR 97214–0450
503–731–4010

Wastewater
Water Quality Division
Dept. of Environmental Quality
811 S. West 6th Ave.
Portland, OR 97204–1390
503–229–5696

Pennsylvania
Drinking Water
Bureau of Water Supply
Division of Drinking Water
Management
Dept. of Environmental
Resources
P.O. Box 8467
Harrisburg, PA 17105–8467
717–787–5017

Wastewater
Bureau of Water Quality
Management
Division of Permits and
Compliance
P.O. Box 8465
Harrisburg, PA 17105–8465
717–787–8184

A - 217

Rhode Island
Drinking Water
Division of Drinking Water
Quality
Dept. of Health
3 Capitol Hill, Room 209
Providence, RI 02908
401-277-6867

Wastewater
Dept. of Environmental
Management
Division of Groundwater and
ISDS
291 Promenade St.
Providence, RI 02908-5767
401-277-2306

South Carolina
Drinking Water
Bureau of Drinking Water
Dept. of Health and
Environmental Control
Bureau of Drinking Water
2600 Bull Street
Columbia, SC 29201
803-7345300

Wastewater
Water Poluution Control
Dept. of Health and
Environmental Control
2600 Bull Street
Columbia, SC 29201
803-734-5300

South Dakota
Drinking Water
Drinking Water Program
Division of Environmental
Regulations
Dept. of Environment and
Natural Resources
Joe Foss Building,
523 East Capitol
Pierre, SD 57501-3181
605-773-3754
800-GET-DENR

Wastewater
Point Source Office
Division of Environmental
Regulations
Dept. of Environment and
Natural Resources
Joe Foss Building,
523 East Capitol
Pierre, SD 57501-3181
605-773-3351
800-GET-DENR

Tennessee

Drinking Water
Dept. of Environment and
Conservation
21st Floor L&C Tower
401 Church St.
Nashville, TN 37243–1540
615–532–0191

Wastewater (Septic Systems)
Groundwater Protection
Division of Water Supply
Dept. of Environment and
Conservation
L&C Tower, 6th Floor
401 Church Street
Nashville, TN 37243-1549
615-532-0191

Wastewater (Sewer Systems)
Groundwater Protection
Division of Water Supply
Dept. of Environment and
Conservation
L&C Tower, 10th Floor
401 Church Street
Nashville, TN 37243-1549
615-532-0761

Texas

Drinking Water
Water Utilities Division
Monitoring and Enforcement
TNRCC
P.O. Box 13087
Austin, TX 78711–3087
512–239–6020

Wastewater
Water Planning and
Assessment Division
Watershed Management
TNRCC
P.O. Box 13087
Austin, TX 78711–3087
512–239–4400

Utah

Drinking Water
Division of Drinking Water
Dept. of Environmental Quality
P.O. Box 144830
Salt Lake City, UT 84114–4830
801–536–4200

Wastewater
Dept. of Environmental Quality
Division of Water Quality
P.O. Box 144870
Salt Lake City, UT 84114–4870
801–538–6146

Vermont

Drinking Water
Water Supply Division
Agency of Natural Resources
Dept. of Environmental
Conservation
103 S. Main Street
Sewing Building
Waterbury, VT 05676–0405
802–241–3400

Wastewater
Wastewater Management
Division
Agency of Natural Resources
Dept. of Environmental
Conservation
103 S. Main Street
Sewing Building
Waterbury, VT 05676–0405
802–241–3822

Virginia

Drinking Water
Dept. of Health
Office of Water Programs
1500 E. Main, Suite 109
Richmond, VA 23219
804–786–6278

Wastewater
Dept. of Health
Office of Environmental Health
1500 E. Main, Suite 117
Richmond, VA 23219
804–786–3559

Washington

Drinking Water
Division of Drinking Water
Dept. of Health
P.O. Box 47822
Olympia, WA 98504–7822
360–753–3466

Wastewater
Water Program
Dept. of Ecology
P.O. Box 47600
Olympia, WA 98504-7600
360-407-6541

West Virginia

Drinking Water
Office of Environmental Health
Services
Engineering Division
815 Cuarrier St., Suite 418
Charleston, WV 25301–2616
304–558–2981

Wastewater
South Charleston District
Office
Office of Environmental Health
Services
1011–A F Street
South Charleston, WV 25303
304–558–0624

Wisconsin
Drinking Water
Bureau of Water Supply
(WS/2)
Dept. of Natural Resources
P.O. Box 7921
101 South Webster
Madison, WI 53703-7901
608–266–0821

Wastewater
Bureau of Wastewater
(WW/2)
Dept. of Natural Resources
P.O. Box 7921
101 South Webster
Madison, WI 53703
608–267–7694

Wyoming
Drinking Water for
Wyoming is handled by:
U.S. Environmental Protection
Agency — Region 8
Drinking Water Branch
(8WM-DW)
999 18th St., Suite 500
Denver, CO 80202–2466
303–293–1716

Wastewater
Water Quality Division
Dept. of Environmental Quality
Herschler Bldg. 4W
122 West 25th Street
Cheyenne, WY 82002
307–777–7781

APPENDIX B

CASE STUDY: WESTERN MARYLAND COOPERATIVE UTILITIES VENTURE

This case study details the work of the Small Towns Environment Program (STEP) with the State of Maryland's Department of the Environment (MDE), the Washington County Sanitary District (WCSD) and six towns in that county: Boonsboro, Clear Spring, Funkstown, Hancock, Smithsburg and Williamsport. Their achievement demonstrates that partnerships can be of benefit to all: to the state, the cooperating entities, and best of all, to taxpayers.

In January 1993 STEP approached MDE officials about using partnerships to achieve small town viability. MDE was receptive to the idea. The state had already invested a considerable amount of money in equipment for preventive maintenance in small systems; it had found, however, that many items were underutilized by the recipient and rarely offered to any other unit of government. MDE recognized that it made sense to demonstrate a model whereby the sharing of equipment was made both feasible and effective.

STEP and MDE therefore sought a pilot project in which participants were not only enthusiastic but also known as innovators. The Washington County Sanitary District and the surrounding rural communities were eager to investigate how such a partnership might work. (See the end of this Appendix for the chronology of this project.)

The group named itself the Western Maryland Cooperative Utilities Venture or WMCUV. Early on, members wished to target a number of diverse issues, but with further discussion agreed to focus on equipment for dealing with inflow and infiltration (I&I), a severe problem all suffered but none could correct by themselves. In prioritizing their wish list, WMCUV members settled on a TV truck, a portable pump station, open channel flowmeter, portable flowmeter, safety equipment, atmospheric monitors, a trailer, and a smoke test machine.

Each town formally resolved to explore collaboration, naming an official representative to the committee's monthly meetings. Both STEP and MDE also attended to guide participants in identifying and addressing basic questions. From the beginning, each town was urged to base decisions on their own needs and problems — not altruism — for long– term support and success.

WMCUV is a useful model of how an enlightened state, Maryland, saw it was to the **state's** advantage to offer a one–time grant for a partnership's start–up costs: $150,000. And the Washington County Sanitary District, a participant, saw it was to **their** advantage to grant WMCUV $15,200 for the state match and to lend an additional $18,000 to be repaid from rental fees. Other participants were therefore not required to contribute any money up front, paying only when they used the equipment, thus assuring greater local acceptance.

THE PARTICIPATORY AGREEMENT

In order to determine WMCUV's basic structure and function, members dealt with a broad array of questions, such as:

Authority:
- Does the state permit such groups, and if so, by what law?
- Is this to be a not–for–profit organization?
- What authority do the partners have – and not have?

WMCUV was established as a non–profit cooperative venture pursuant to the ***Annotated Code of Maryland*** and the Charters of the member towns. It can purchase and lease equipment, contract for services, and set rates and policies. It does not have authority to impose taxes, pledge the credit of its members, borrow money or purchase goods or services on credit, acquire any interest in real estate, or build or operate either a water or a wastewater system.

Mission:
- What is the purpose of the partnership?
- What area will it serve?
- Is it limited to wastewater?
    - If so, just to the collection system?
    - To just operation and maintenance of the system?
    - To share management services for the system?
    - Exactly what levels of service will be provided?

WMCUV's stated mission is to "make use of opportunities to share equipment, services and resources among participating governmental entities to maximize the efficiency and effectiveness of local operations and preventative maintenance of wastewater systems."

Governance:
- Are all participants equal?
- Who will represent each government, and how will they be chosen?
- May other partners be added — and may present partners withdraw or be expelled?
- If a separate governing body is established, what are their voting rights?
- How many votes constitute a quorum, and are proxy votes permitted?

- How often will they meet?
- Regarding administration:
  ○ Who makes day–to–day decisions?
  ○ How will that person be chosen?
  ○ How long will he/she serve?

WMCUV established an Executive Board composed of officially–designated representatives of the members. The Board, meeting at least quarterly, is responsible for establishing policies and for annual appointment of an administrator. Each member has one vote regardless of size (this provision addressed an early concern that the WCSD, being much larger than the others, might become dominant). Four votes are required for positive action. The Board may expel a member for failure to pay properly imposed charges, and upon 60 days written notice, any member may withdraw.

The first year's Chairperson was Julianna Albowicz, Vice–Mayor of the Town of Clear Spring; the first Administrator was Lynn Palmer, who was also Executive Director of the WCSD. MDE is not a party to the agreement, but as technical advisor will have access to all meetings, correspondence, financial records, and reports. STEP's role ended with the start of WMCUV's implementation.

General provisions:
- How will amendments to the document be handled? (The process should be specified.)
- Is there to be a periodic review? (Annually is standard.)

A town's attorney can recommend language on severability, dispute resolution, incontestability, and holding harmless. Other municipal documents might contain these standard provisions, too, which could be copied. Additional references might be your Secretary of State or New York State's *Intergovernmental Co-*

operation cited previously. This 12–page booklet gives overall considerations plus illustrative contract clauses that are very helpful. You can order a copy by writing to the following address:

> Office for Local Government Services
> NYS Department of State
> 162 Washington Avenue
> Albany, NY 12231

POLICIES

The participatory agreement can be thought of as the group's constitution: the principles which should endure for some time. A constitution should be amended as needed, but only after due process and formal adoption. Policies, on the other hand, may require more frequent change to reflect the knowledge accumulated as the partnership gains experience. For partnerships that deal with shared equipment, some policies may include:

Operation and maintenance
- What are the detailed elements of the administrator's job?
- Does he/she receive a fee or salary?

In WMCUV, the administrator was responsible for implementing Board policies; purchasing, procuring and financing equipment, materials, supplies and services; establishing and maintaining financial records; and scheduling and supervising the use of equipment, materials and supplies.

- How will breakdowns from normal wear be handled?
- What provision will there be for routine maintenance?

The WCSD stores and maintains WMCUV's equipment as each piece is returned, thereby avoiding preventable breakdowns at the work site. Customers agree to pay the current list price for damage or loss regardless of cause, except for reasonable wear and tear, while the goods are out of WMCUV's possession.

- What are the terms and conditions of use, e.g., prohibited uses; penalties for dirty, damaged or lost equipment; time of return, time of payment, collection costs, and repossession?

WMCUV has detailed policies on all these matters; they are made explicit to borrowers in the rental contract.

Scheduling
- Who has priority for usage?
- How is scheduling arranged and controlled? Who makes those decisions – and is there an appeal?
- Is there special provision for emergencies?

The WMCUV administrator is responsible for making common-sense decisions, based on priority of need. A member may appeal to the Board.

Rate structure
- If costs are to be covered by rates, are these charges uniform for all participants?
- If not, what are the specifics for those with differing consumption or for outside customers?
- What elements of the operation do the rates cover, e.g. maintenance, debt retirement, depreciation, replacement costs, insurance, etc. ?

The WMCUV Board established the rate structure, providing for three distinct tiers of charges. The lowest rates are paid by original members, recognizing their contribution of time and effort to establish the partnership. Other towns in Maryland pay the next higher level of rates, benefitting from the state's investment in WMCUV for broader use. The highest costs go to entities in the nearby states of Pennsylvania and Virginia and commercial operations since they have no investment in WMCUV at all.

The following table gives a comparison of rental costs for one of the main pieces of equipment procured by WMCUV:

Cost-per-foot Comparisons for TV Truck			
	WMCUV*	Outside Contractor A**	Outside Contractor B
Hourly rate	$ .63	$1.00***	$1.70***
Daily rate	.38		
Weekly rate	.26		
Monthly rate	.22		

*   Rates are based on projected usage as of 6/15/94.
** Requires a minimum of 1000 ft.
*** These are flat fees charged for all terms of usage.

The difference between rates offered to partners by WMCUV and those charged by outside firms is substantial. Based on projected usage, members pay from $.22/ft. for the monthly rate to $.63/ft. for the hourly rate. The outside contractors, on the other hand, charge flat fees ranging from $1.00/ft. to $1.70/ft. WMCUV members enjoy not only substantial savings but convenience: six items of their jointly–owned equipment are not available for local rental at any price.

Rental rates include all the usual factors discussed above, plus operating costs based on actual usage, not limited to labor, gas, oil, materials, and supplies.

Procurement Policies
- What amounts permit telephone quotes, written quotes and formal bids?
- Does it matter if items are new, emergency purchases, or replacements?

Again, these matters are covered in WMCUV's Procurement Procedures.

Liability
- How is liability determined?
- Will sufficient insurance coverage be in place from the beginning?

Insurance is a requirement even before the equipment leaves the yard. Coverage of WMCUV property was supplemented by policies maintained by each municipality.

Recommendations:
In applying this model to other clusters of communities, we would make several recommendations about the structure of the exploration: the group should be small enough to be efficient, representative enough to reflect most local opinion, and skilled enough to raise useful questions. More specifically:

- The group should number no more than nine, and no fewer than three people who are really motivated to find answers.
- If all participants are government officials, they should reflect concerns of the broader constituency.
- The group should include some people with especially appropriate knowledge or skills, such as an attorney, an accountant, or a contractor.

The funding of the partnership is an essential consideration. Even if participating towns begin with a handshake, they need to anticipate expenses for the arrangement, such as maintenance, repairs, transport, telephone, insurance, etc., plus the time of personnel involved. It is not enough to merely hope that the partnership will be financially self- sufficient. Somebody needs to look at projected costs and revenues for reasonable expectations of what is **likely**.

In other cases, partnership agreements have been either formal or informal, oral or written. The casual arrangement may work well under some circumstances, but with WMCUV we believed strongly that a document that anticipated implementation issues and defined remedies for disputes was well worth the time and care necessary to prepare it.

Before the member governments and their voters were persuaded to approve the arrangement, various briefings and draft documents were submitted for their consideration. For both present and future reference, all documents were as succinct as possible without sacrificing clarity, and in language that was easily understood.

Consistent with STEP's practices, all available WMCUV participants analyzed their own lessons learned as operations were about to begin:

On the initial incentive to explore a partnership
- **For the state:**
  STEP's approach provided the spark to do what MDE had been thinking about, i.e., finding a way to use its grant money more efficiently. MDE realized it was making some grants ineffectively, and with limited resources it wanted to find a way to make its money go farther.

- **For the Sanitary District:**
    - They wanted a TV truck.
    - This was a way to spread fixed costs.

- **For the towns,** a complex variety of inducements got them to the table:
    - Social reinforcement: Personal invitation from a person with status.
    - Money: This might be an opportunity to get a state grant.
    - Public responsibility: The money would probably go for maintenance equipment.
    - Innovation: Towns saw the need, but didn't know what to do about it.
    - Local solidarity: It was an opportunity to work with other towns, sharing problems and ideas.
    - Local control: This might be a way to preserve their independence from the largest city nearby and avoid dominance by the Sanitary District.
    - Shared risk–taking: The joint effort gave protective cover to all despite some personality clashes as well as concern about not knowing the towns' responsibilities.

Members' assessment of WMCUV's creation
- Member governments now provided better service to taxpayers while saving money.
- Member towns had cohesiveness. The towns were all from the same county and representatives knew each other.
- Towns moved to deal with a demonstrated need. It cost the Sanitary District $50,000 for just three winter months of 1993-94 to treat the wastewater of one subdistrict due to severe I & I problems.
- Members are glad they started with a limited focus. "Otherwise, we would have talked about everything and solved

nothing," one said.
- Developing all the documents collaboratively was a positive process:
    - It enabled members to see progress – and reassure their mayor and council.
    - It provided security, demonstrating that they would retain a voice.
    - It enabled mutual growth and learning about issues and technology.
    - It developed confidence that the towns could establish and operate the venture by themselves.
- The state's encouragement and support was unexpected but very gratefully received. The towns had feared that the state's participation would be heavy-handed and strongly prescriptive when in actuality the opposite was true.
- This venture would not have been possible without the Sanitary District's providing the match to the state grants. The towns would have been unable to come up with the money — although STEP could have helped them secure other funds had that been the case.
- While everyone was aware of the financial match, there was little acknowledgment of the time, travel and preparation time contributed by each of the towns.

Outcomes
- There is a significantly improved understanding of the benefits of a partnership. One said, "We could now go on to other issues. Everybody needs a brush chipper — and a street sweeper!"
- There is much greater intergovernmental confidence. Member towns would probably now be willing to contribute start-up money for another venture, based on the successful result of the present one.
- The attention directed to the I&I problem has increased

local awareness of the need to **plan** for maintenance expenditures, rather than waiting for emergencies to arise before acting. In 1994 one town set aside $15,000 for I&I remedial work, the first time such a line item ever appeared in its budget.

Recommendations for replication
- Seek clusters of towns that have a common problem.
- Select representatives who are motivated to find answers and will stick with it until the solution is implemented.
- Show others the WMCUV model, but allow them to adapt and modify it as needed.
- Help them to understand this is an **investment** not only in equipment but in local self–reliance. The process may involve some education of local governments as well as their constituents, emphasizing the value of preventive maintenance.
- Make sure there is a local sparkplug: an energetic, capable person who will spearhead the local effort. It's **people,** not plans or documents, who determine feasibility.
- Select a narrow mission. It can be expanded later.
- Get outside guidance from people who know government, finance, municipal law, and how to write appropriate specifications.
- Develop the right structure and agreements so the venture becomes permanent. Anticipate problems of implementation so answers can be written and ready.
- Develop a marketing plan, including publicity, to generate income immediately.

While it may not be until fall of 1995 that WMCUV has performance data and cost comparisons with business–as–usual, MDE, WMCUV and STEP began looking for other applications of this model a year earlier. WMCUV demonstrated to all participants that very small towns **can** collaborate effectively while protecting their own self–interest and improving their environment.

WMCUV CHRONOLOGY

1993

January –
- STEP approached MDE about partnerships to achieve small town viability. MDE liked the idea, inviting Lynn Palmer of the Washington County Sanitary District to participate.

February –
- A meeting was held in Angelo Bianca's MDE office in Baltimore with George Keller of MDE, Jane Schautz of STEP, Mr. Palmer, and representatives of 4 other towns: Boonsboro, Hancock, Smithsburg and Williamsport.
- Agreement was reached to focus on only I&I to start.
- MDE would invest start–up seed money if the partnership would sustain O & M. Therefore, members could pay only for the equipment they use.
- Questions were raised whether local mayors and councils would join when costs were not known.
- Participants agreed to prepare a wish list of equipment for the next meeting and think about organizational structure.
- By the end of the month, all towns agreed to explore the partnership and appointed an official representative.
- Schautz sent Keller an outline of topics for the participatory agreement.

March –
- The meeting was held in Washington Co. at the Sanitary District with Clear Spring now represented, also.
- WCSD provided the $15,000 match for MDE's capital grant of $125,000.
- Keller drafted a list of options for shared equipment, expressing the state's interest in long–term equipment vs. the locals' preference for emergency items. He also presented options for institutional arrangements, legal arrangements, and implementation.
- Discussion was held on how the state's subsidy would be spread: 100% of largest items or lesser percentage of more items?
- The partnership name (with pronounceable acronym) was decided: Western Maryland Cooperative Utilities Venture (WMCUV).
- The group chose priority purchases: TV truck, portable pump station, open channel flowmeter, portable flowmeter, safety equipment, atmospheric monitor, trailer, smoke test machine.

April –
- Keller presented a draft statement of the kinds of agreements required.
- Towns needed to pass a formal resolution to participate. Representatives have been keeping Councils informed.
- Issues still to be resolved: scope and expansion of membership, structure/function of the governing board, mission statement, financing/rate structure, purchasing policies, administrative duties, rules for use, complaints and remedies, legal advice, who owns the equipment: WMCUV or WCSD?

May –
- Four of six members executed an agreement of intent to participate: Boonsboro, Clear Spring, Smithsburg, WCSD.
- Detailed discussion was held on authority of WMCUV, its mission, governance, and general administration.
- Grant application to MDE was filled out.

June –
- WCSD's attorney advised that state law provided for the establishment of a non–profit, quasi–public organization. There was no need to invent a new structure.
- Group agreed on 3 levels of rates: lowest for members, higher for the rest of MD, highest for out–of–state.
- Keller agreed to draft language on policies for the next meeting.

July –
- Contracts were drawn up for submission to MDE.
- The final version of participatory agreement was sent to all members for discussion and approval.

August –
- Boonsboro, Clear Spring, Smithsburg, and WCSD signed the participatory agreement without reservations.
- Hancock signed, but asked for clarifications on questions/concerns.
- Keller wrote mayors of both Hancock and Williamsport answering questions/objections.
- Discussion was started on policies: rate structure, etc.
- Julianna Albowicz was elected as Chairperson, Lynn Palmer as the unpaid Administrator – both for one year.

September –
- Williamsport signed the agreement on September 7.
- Keller announced approval by the Public Utility Board of the grant of $125,000.
- Formal bids for equipment opened. Some items needed to be re–bid.
- Policies discussed included financing, purchasing, priority of use, insurance, etc.

October –
- Contracts were awarded for everything that was affordable.
- MDE might grant an additional $25,000 to purchase more items; WCSD would lend $18,000 to be repaid through rental fees.
- Preliminary rates show a 50% rate reduction for members' use of WMCUV equipment.
- Vance Isherman of Williamsport was appointed Treasurer.

November –
- Funkstown attended meeting.
- Tier I (members) rates adopted with recommendation that Tier II be 10%–15% higher.

1994

January –
- Three others towns were invited to join but did not respond.

February –
- The Town of Funkstown was accepted as a member.
- A bank account was established.
- Change orders were approved on some equipment procurement.

- Approval was given to purchase the remaining equipment.
- Keller presented a draft rental contract application form for discussion.

April –
- Insurance options discussed included workers compensation.
- Legal issues discussed included incorporation for limiting liability, tax implications, conflict with local charters, e.g., purchasing, ownership of equipment, etc.
- The TV truck was delivered.
- The gas monitoring equipment arrived, too, with training attended by Boonsboro, Williamsport and WCSD.

May –
- Training was held on the TV truck.

June –
- Insurance coverage was determined: inland marine, business auto (for the van and trailer), and general liability.
- Rates for Tier II were established at 15% more and Tier III at 30% more than those for Tier I.
- A marketing strategy was developed: descriptive brochures, flyers, articles in newsletters, etc.
- An attorney was authorized to initiate incorporation proceedings.
- A lessons learned session was conducted.

July –
- A dedication ceremony was held, attended by the County Commissioners, state legislators and the media.

APPENDIX C

ON MONEY... and knowing how much you can afford to borrow

When doing self–help projects, it's best to begin with what people can afford, then try design the house, water system, or other needed improvement with that limit in mind. The chart on page C–3 is a tool for helping to determine what a borrower (whether a household seeking a home or a community seeking a water line) can afford as a loan limit given their ability to repay.

There are two ways to use this chart.

First, you can determine the monthly payment on a given loan. Assume you are borrowing $75,000 @ 10% interest for 15 years. Look down the column labeled "15 Years" until you get to the number that corresponds to 10% interest. It is 10.7461. Now multiply this factor by the number of thousands in $75,000 (10.7461 x 75 = $805.9575). You have now established your monthly payment of about $806 as needed to cover both principal and interest on a loan of $75,000 if the terms are 10% interest for 15 years. If you want the annual payment, just multiply by 12 and get $9,672. If you are in the mood to practice, try the monthly cost of repaying a loan of $150,000 @ 7% with a 30–year term. Is your answer $998? You got it!

Second, start with what people can pay, then establish how much they can borrow. Assume 45 households in a community can each afford $12 per month for a distribution system that will bring good water to their tap. How much can they afford to borrow? Forty– five payers x 12 gives us a monthly payment potential of $540. Let's assume loan terms of 8% interest for 25 years— which gives us a factor from the chart of 7.7182. If we divide the $540 by this factor, we get 69.964, carrying it to the nearest three places. Now, replace the period with a comma and get $69,964 as the total they can borrow with this payment ability. If you want to reverse the logic, note that there are 69.964 thousands in this total. Multiply this by the factor (7.7182) to get $540.

If you want more practice, try a monthly payment ability of $480 and assume conditions of a 25-year term at 6% interest. If your calculation of how much can be borrowed is $74,500, you're right.

Note that these calculations do not include closing costs (which may increase the amount borrowed) or the effect of a down payment (which will decrease the amount borrowed).

This tool can also be used to help you bootstrap by lowering both the cost of the project and the cost of borrowing. Let's say you start with a payment ability of $350 per month (for principal and interest), and a projected cost for the system you want to build of $55,000.

We begin by finding out what can be borrowed through a conventional 10% loan for 20 years. With a $350 payment, the answer is $36,269. Now we know what our problem is: the "shortfall" of about $17,000 between the loan you need and the loan you can afford.

Our first approach is cost reduction. Let's say we use voluntarism and other factors to reduce the cost to $45,000. How do we cover the remaining $7,000 shortfall? One way is to use our chart to see if a better loan deal would let this happen. If we project a 30-year term at 8%, the amount this community can borrow rises to $47,700. We're there! And we have a strong rationale for approaching the bank or other borrower. Rather than asking, "How low can you go?", we can specify just what we need to make the deal work.

With a combination of lowered amount needed and better loan terms, many impossible deals are made possible.

Table of Loan Factors

Interest Rate	Monthly Payment in Dollars Per $1,000 of Principal			
	15 Years	20 Years	25 Years	30 Years
0%	5.5556	4.1667	3.3333	2.7778
1%	5.9849	4.5989	3.7687	3.2164
2%	6.4351	5.0588	4.2385	3.6962
3%	6.9058	5.5460	4.7421	4.2160
4%	7.3969	6.0598	5.2784	4.7742
5%	7.9079	6.5996	5.8459	5.3682
6%	8.4386	7.1643	6.4430	5.9955
7%	8.9883	7.7530	7.0678	6.6530
8%	9.5565	8.3644	7.7182	7.3376
9%	10.1427	8.9973	8.3920	8.0462
10%	10.7461	9.6502	9.0870	8.7757
11%	11.3660	10.3219	9.8011	9.5232
12%	12.0017	11.0109	10.5322	10.2861
13%	12.6524	11.7158	11.2784	11.0620
14%	13.3174	12.4352	12.0376	11.8487
15%	13.9959	13.1679	12.8083	12.6444
16%	14.6870	13.9126	13.5889	13.4476
17%	15.3900	14.6680	14.3780	14.2568
18%	16.1042	15.4331	15.1743	15.0709
19%	16.8288	16.2068	15.9768	15.8889
20%	17.5630	16.9882	16.7845	16.7102

NOTE: These are general factors. The actual amounts charged by a lending institution may vary slightly in either direction.

APPENDIX D

Washington State Department of Ecology
RECIPIENT GUIDANCE IN-KIND POLICY
FOR SMALL TOWNS ENVIRONMENT PROGRAM
(STEP) PROJECTS

Water Quality Financial Assistance will allow in-kind for facilities involved in STEP. The following decisions on how to implement the in-kind policy have been made.

Implementation Steps.

- In-kind will only be allowed on facility-type projects that are selected to be in STEP.

- Ecology will not track individual efforts but will negotiate up front with the community the value (or worth) of a particular task. The project officer and project engineer will use available tools to negotiate the worth of the task.

- The recipient will fill out a work task form (attached) which outlines the task, how the task will be completed and estimated value. This will be the basis for negotiation with Ecology.

- When the task is complete to Ecology's satisfaction, the project officer will allow the negotiated worth as match. This will eliminate the need to track individual timesheets. If it takes longer for volunteers to do the task or requires some training of the volunteers, we will only allow the negotiated worth of the task.

- Ecology will not pay the Town cash for the volunteer labor. It will be allowed as match only. If available, grant/loan funds can be used for materials/equipment directly utilized on the project.

- If the volunteer or force account task was poorly completed and unacceptable to Ecology, the cost of rework will be ineligible for grant/loan participation.

- The utilization of in–kind will be addressed on a case–by–case basis. Using in–kind will vary from project to project given the amount of expertise and results available in a community. Ecology staff will assist the communities in determining the level of in–kind reasonable for the project.

- To be approved, in–kind must be less expensive or at least no more expensive than conventional contracting methods as estimated.

D - 245

WORK TASK FORM FOR UTILIZING IN-KIND SERVICES

Recipient: _____   _____
Task No. _____   _____
Description of task to be completed: _____   _____   _____   _____
Plan for completion of task: (List who will be involved, tools/equipment needed, timeline for completion.)   _____   _____   _____   _____   _____   _____   _____
Estimated value of task: _____   _____   _____
How did you estimate the value?   _____   _____   _____   _____   _____   _____   _____
Recipient signature: _____   **Date** _____
Department of Ecology   **Approval signature:** _____   **Date** _____

APPENDIX E

Request for Proposals

Village of Massena, New York
Wastewater Treatment System

February 1994

Hassan A. Fayad, *Superintendent of Public Works*
Michael Weil, *Village Engineer*
Steve Siddon, *Plant Superintendent*

Authors' note: The following is excerpted from Massena's RFP of February, 1994. Of particular interest are those sections which specify or describe what is expected of the engineer in terms of the content of the proposal as well as his/her role in the community's self-help wastewater project:

Pages	Section	Topic
E-256	1-G	Inclusion of staff comments and suggestions.
E-258 E-259	1-L	Local involvement in preproposal schedule
E-261	1-P	Presence of the project engineer at interview with the Village.
E-262	1-Q	Providing all supporting information used to create Final Design Report.
E-263	2-A	Village's general stance regarding affordability and self-help.
E-264	2-B.c	Attendance at meetings with Village Board, sewer committee, and the public.
E-265	2-B.l	Providing cost estimates for conventional approach.
E-265	2-B.m	Identification of tasks that can be done by local forces, as well as cost estimates for self-help approach.
E-266	2-B.t	Providing self-help analysis.
E-270	3-C	Identification of self-help tasks.
E-271	3-D	Description of previous self-help experience.

TABLE OF CONTENTS

PROJECT INFORMATION

Part 1 GENERAL
- A. PREFACE — E-252
- B. DESCRIPTION OF VILLAGE — E-252
- C. SYSTEM DESCRIPTION — E-253
- D. CURRENT OPERATING PROBLEMS — E-254
- E. PROPOSED PRELIMINARY PROJECT SCHEDULE — E-254
- F. AVAILABLE INFORMATION — E-255
- G. ENGINEERING APPROACH — E-256
- H. FUNDING — E-256
- I. CONTRACT COMPENSATION — E-257
- J. REJECTION OF PROPOSALS — E-257
- K. INCURRING COSTS — E-258
- L. PRE–PROPOSAL SCHEDULE — E-258
- M. RECEIPT OF PROPOSALS — E-260
- N. PROPOSAL FORMAT — E-261
- O. DISCLOSURE OF PROPOSAL CONTENT — E-261
- P. VERBAL PRESENTATION — E-261
- Q. SUPPORTING INFORMATION — E-262

PART 2 DESCRIPTION OF PROFESSIONAL SERVICES
- A. GENERAL — E-263
- B. PRELIMINARY ENGINEERING PHASE — E-263
- C. FINAL DESIGN PHASE — E-266
- D. CONSTRUCTION ASSISTANCE — E-268

PART 3 INFORMATION REQUIRED FROM PROPOSERS
- A. STATEMENT OF THE PROBLEM — E-270

B. MANAGEMENT SUMMARY	E-270
C. WORK PLAN	E-270
D. PRIOR EXPERIENCE	E-271
E. PERSONNEL	E-272
F. INSURANCE	E-272
G. COST AND PRICE ANALYSIS	E-272

PROJECT INFORMATION

ISSUING AGENCY:
VILLAGE OF MASSENA
DEPARTMENT OF PUBLIC WORKS
WASTEWATER TREATMENT PLANT
536 SOUTH MAIN STREET
MASSENA NY 13662

MUNICIPAL OFFICIALS:
CHARLES R. BOOTS, *Mayor*
JOHN R. FEELEY, *Trustee*
WAYNE E. LASHOMB, *Trustee*
CLIFFORD LITTLEJOHN, *Trustee*
LAWRENCE I. PRASHAW, *Trustee*
SANDRA A. SMITH, *Clerk*
DANIEL E. CASE, *Treasurer*
RANDY L. PEETS, Esq., *Attorney*

ISSUE DATE:
FEBRUARY 15, 1994

RESPONSE DEADLINE:
MARCH 29, 1994; 2:00 P.M.

SEWER COMMITTEE:
CLIFFORD LITTLEJOHN, *Trustee*
JOHN R. FEELEY, *Trustee*
HASSAN A. FAYAD, *Superintendent of Public Works*
STEVE SIDDON, *Plant Superintendent*
MICHAEL WEIL, *Village Engineer*

CONTACT PERSON:
STEVE SIDDON, *Plant Superintendent*
VILLAGE OF MASSENA
536 SOUTH MAIN STREET
MASSENA NY 13662
(315) 764–0653
Fax   764–9948

REQUEST FOR PROPOSALS
VILLAGE OF MASSENA
NEW YORK

PART 1　GENERAL

1-A.　PREFACE

The Village of Massena is requesting New York State licensed engineering firms to present proposals for preliminary engineering services. The services will be for additions, upgrades, modifications, and/or deletions of existing treatment components that serve the Village and a portion of the Town of Massena. Consideration will also be given for a new plant at either the existing site or a new site. The new design must be capable of producing effluent which will meet the anticipated SPDES limits. The design capacity will include an increased flow based on future growth projections. The final design capacity and SPDES limits will be determined during the preliminary engineering phase of the project.

The Village is also seeking cost estimates for design and construction services. The Village will receive proposals until March 29, 1994 at 2:00 pm at the office of the Superintendent of Public Works, 536 South Main Street, Massena, New York, 13662.

Proposers attention is directed specifically to *Part 1-L* regarding plant visits and to *Part 3-G* regarding proposal submission.

1-B.　DESCRIPTION OF VILLAGE

The Village of Massena is located in northern St. Lawrence County approximately 3 miles south of the St. Lawrence River and the Canadian border. The current Village population is about 12,000 and does not experience wide seasonal fluctuations. The economy of the Village is primarily retail and light commercial. Heavy commercial and manufacturing industries are located in the Town of Massena and are not serviced by the Village wastewater system.

1-C. SYSTEM DESCRIPTION

From 1959, when the original wastewater treatment plant was constructed, until 1980, the plant provided primary treatment of wastewater discharged to the Grasse River. In 1980 the plant was upgraded to provide secondary treatment. This design was unable to meet the SPDES requirements until the addition of a polymer system. The plant now produces an acceptable but expensive effluent.

The wastewater collection system primarily serves the Village of Massena, including a small section of the Village located in the Town of Louisville. It also serves several areas in the Town of Massena including a small sewer district south of the Village as well as two existing sewer districts east of the Village. The easterly districts encompass the Highland road area, the St. Lawrence Centre Mall, and the new Massena Towne Centre development.

The original network of sanitary sewers was constructed during the 1880's. Excluding service laterals, the collection system currently consists of approximately 48 miles of concrete, asbestos cement, clay tile, and PVC pipe which vary in size from 4" to 42". The system is comprised mainly of gravity sewers with six lift stations conveying wastewater from outlying areas. Three river crossings and approximately 800 manholes connect the system.

1-D. CURRENT OPERATING PROBLEMS

The wastewater treatment plant has numerous operating problems. The DEC has noted several of the most severe problems the plant encounters in their recent inspection reports. Some of these include exceeding our permit flow limit, obsolete and worn equipment, specialized expensive repair parts leading to excessive down time, and excessive objectionable odors caused by solids accumulation.

The collection system problems identified for correction under this program include pump station rehabilitation and investigation of the primary pump station receiving trunk sewer. Since the mid 1970's, the Village has spent considerable monies in the collection system to reduce the amount of I/I entering the system. The Village has been successful in substantially reducing inflow. Infiltration cannot be effectively removed.

1-E. PROPOSED PRELIMINARY PROJECT SCHEDULE

    a.) Issuance of RFP Feb 1994
    b.) Informational Meetings Mar 1994
            See *Part 1.L* for detailed schedule)
    c.) Receive and review proposals Mar 1994
    d.) Select 3 proposals for Final Review Apr 1994
    e.) Conduct interviews Apr 1994
    f.) Selection of Engineer Apr 1994
    g.) Negotiate and execute contract Apr 1994
    h.) Preliminary Report Aug 1994
    i.) SEQR Aug 1994
    j.) Final Engineering Report Sep 1994
    k.) Advertise for Final Design Engineer Sep 1994
    l.) Selection of Engineer Nov 1994
    m.) Complete Final Design Mar 1995

n.) Receive Bids for Construction May 1995
o.) Begin Construction Jun 1995
p.) Complete Construction Jun 1996
q.) Start Up of New Plant Jul 1996

1-F. AVAILABLE INFORMATION

Information is available for review at the Village of Massena Wastewater Treatment Plant from 7:00 am to 4:00 pm from March 1 – March 10. Please call Steve Siddon at (315) 764–0653 in advance to arrange for access to the information. Proposers are strongly urged to review this material. It will be assumed that all proposals have taken cognizance of and reflect the relevant data and information contained in such materials. The Village will make the following information available for inspection to all proposers:

a.) Plans of the Wastewater Plant, 1959 and 1980
b.) Sewer system maps
c.) County tax maps
d.) Pertinent DEC correspondence
e.) DMR reports
f.) VHS sewer inspection tapes/logs
g.) June 1969 Lozier Engineering Report
h.) February 1975 Lozier Engineering Report, Addendum No. 3
i.) November 1975 Lozier Engineering Report, Addendum No. 4
j.) February 1980 Lozier Infiltration Flow Isolation
k.) January 16, 1992 Tisdel Associates Engineering Report
l.) October 1993 Improvements Report

The above information was not specifically designed for and is not a part of this RFP. The Village shall not be responsible for nor make any representation or guarantees as to accuracy, completeness, or pertinence of the information. In addition, the Village shall not be responsible for any conclusions drawn from the above information.

1-G. ENGINEERING APPROACH

The Massena Village Board, through the Department of Public Works and the Wastewater Treatment Plant has been working towards improving the plant and collection system. The Village desires to have a wastewater system that is technically feasible, cost–effective, and that uses technology proven in other communities with similar technical, financial, and climactic requirements and constraints. The design must take population and flow growth into account. If design for full future growth produce extremely high proposed user rates, alternate progressively staged growth solutions should be investigated and recommended in detail. Accordingly, the Village will examine proposals and place primary emphasis on proposers' approach to solving the problem.

Concurrent with the execution of the Final Design Contract for Engineering Services, the Village will require the selected design engineer to issue a performance guarantee. This guarantee will provide specific assurances to the Village that the upgraded plant will operate as designed.

The Village Board has directed that the selected engineer must include the operational staff's comments and suggestions as an essential part of the design process.

1-H. FUNDING

In order for the Village of Massena to finance a project of this magnitude, the Village is including funding investigation as an integral part of the preliminary engineering scope of work.

The Village has begun investigating the NYS Revolving Fund program as one source for funding. The Village is seeking an engineering firm with a demonstrated ability to secure funding and to minimize project costs.

1-I. CONTRACT COMPENSATION

The intent of this RFP is to procure preliminary engineering services only. The method and amount of compensation shall be negotiated. The Village anticipates a "cost plus fixed fee, not to exceed" contract basis.

The Village prefers to have the same engineering firm provide preliminary engineering, design, and construction services. Through this RFP the Village is seeking estimates for the design and construction services for informational purposes and to help us evaluate proposers qualifications.

The Village intends to proceed with the wastewater collection/treatment project based on the results of the preliminary engineering work. The award of the preliminary engineering service contract as a result of this RFP **does not** represent any commitment that the Village will award a design/construction service contract to any proposer. It is the Village's intention to execute separate contracts for final design and construction services. Before such awards, the Village expects to be positioned to procure funding as a result of the preliminary engineering contract.

1-J. REJECTION OF PROPOSALS

The Village reserves the right to:
- a.) amend, modify, or withdraw this RFP;
- b.) require supplemental statements or information from proposers;
- c.) extend the deadline for responses to this RFP;

d.) reject any or all proposals received pursuant to this RFP;
e.) waive or correct any irregularities in proposals, after prior notice to the proposer; and
f.) negotiate separately with competing proposers.

1-K. INCURRING COSTS

This RFP does not obligate the Village to award a contract, to pay the costs incurred in preparing any proposal, or to procure the services described herein.

All proposals are submitted at the sole cost and expense of the proposer. The Village shall incur no liability or obligation to any proposer except pursuant to a written contract for services, duly executed by the proposer and an authorized signatory for the Village.

1-L. PRE-PROPOSAL SCHEDULE

a.) Considering the number of expected proposals and the amount of information available, plant tours and information distribution will be according to the following schedule:

February 15, 1994 — RFP Issued
March 1, 1994 — Informational Meeting
March 1–10, 1994 — Information Available
March 11, 1994 — Final Informational Meeting
March 16, 1994 — Written Questions Deadline
March 22, 1994 — Written Responses to all Proposers
March 29, 1994 — Proposal Due

b.) **INFORMATIONAL MEETING:** The Village will hold an Informational Meeting on March 1, 1994, 9:00 am at the Wastewater Treatment Plant. The purpose of this meeting is to acquaint all proposers with the physical configuration of the plant and collection system. Although not mandatory, the Village strongly encourages all proposers to attend. The Village will answer questions at that time. A form will be available at the meeting to request copies of selected information. Copy costs will be billed by copy source directly to proposer.

c.) **INFORMATION AVAILABLE:** All proposers will be able to inspect available information and make written requests during the March 1–10 time period. There will be no plant inspections or answers to verbal requests for additional information during this time.

d.) **FINAL INFORMATIONAL MEETING:** The Village will hold a Final Informational Meeting on March 11, 1994, 9:00 am at the Wastewater Treatment Plant. The purpose of this meeting is to answer questions not adequately addressed at the informational meeting and to answer written questions received. As with the first meeting, all proposers are urged to attend.

e.) **WRITTEN QUESTIONS DEADLINE:** Further questions should be submitted by March 16. The Village will give consideration to all questions submitted and will send clarifications to all proposers.

f.) **WRITTEN RESPONSE TO ALL PROPOSERS:** The Village will send written clarifications to all proposers on March 22. The Village will not accept questions after that date.

g.) **PROPOSAL DUE:** Responses are due by the time stated above.

All questions or requests for information should be in writing, and should be directed to Steve Siddon, Plant Superintendent, at the address listed on this page. Verbal requests for additional information will not be honored.

Proposers should be aware that the Village reserves the right to inform all RFP recipients of the answers to questions submitted, or additional information provided. The Village, in its sole discretion, will determine if such answers or information are germane to all potential proposers, or if the answers change or alter any material provision of this RFP. The Village <u>shall be under no obligation</u> to provide such answers or information to proposers who choose not to attend meetings, but will make an effort to inform proposers of important facts not previously included.

1-M. RECEIPT OF PROPOSALS

To be considered, proposals must be received by Hassan A. Fayad, Superintendent of Public Works, Village of Massena, 536 South Main Street, Massena, NY, 13662 by 2:00 pm, March 29, 1994.

Proposals must be submitted in a sealed opaque envelope plainly marked with both:

VILLAGE OF MASSENA
NAME AND WASTEWATER TREATMENT
PLANT <u>and</u> ADDRESS OF PROPOSAL PROPOSER

Proposals received prior to the time of opening will be kept sealed until the stated opening time. Proposals received after such time <u>will not</u> be considered.

When proposals are sent by mail, the sealed proposal,

marked as above, should be enclosed in an additional envelope similarly marked and addressed to the person stipulated above. If mailed, adequate time for its delivery should be allowed. The Village will not be responsible for proposals lost or delayed in the mail.

The Village encourages delivery of proposals by a method which requires a signature to acknowledge receipt.

1 -N. PROPOSAL FORMAT

To be considered, proposers must submit a complete response to this RFP, using the format provided in **Part 3.** Proposers are to submit six (6) copies of their proposal. Proposals must be signed by an official authorized to bind the proposer to its provisions. The proposal must remain valid for at least 90 days. The contents of the proposal of the successful bidder will become part of the basis for contractual obligations if a contract for preliminary engineering is entered into.

1-O. DISCLOSURE OF PROPOSAL CONTENTS

Information provided in your proposal will, to the extent allowed by law, be held in confidence and will not be revealed or discussed with competitors. If a proposal contains any information that the proposer does not want disclosed to the public or used by the Committee for any purpose other than evaluation of the offer, each sheet of such information must be marked **(Confidential)**.

1-P. VERBAL PRESENTATION

Proposers may be invited to make a verbal presentation of their proposal to the Committee. Such presentations provide an opportunity for the proposer to clarify its proposal to the committee in order to insure a thorough understanding of the material submitted. The Committee will schedule all presentations.

The presence of the project engineer who will be working specifically on the project (as opposed to a sales/contracting representative or corporate officer) will be required at the presentation.

1–Q. SUPPORTING INFORMATION

As an addendum to the Final Design Report, the preliminary design engineer must surrender to the Village all supporting information used to make final design determination. Such information includes, but is not limited to:

a.) design calculations;
b.) equipment catalog cut sheets considered;
c.) alternate treatments considered;
d.) cost estimates;
e.) funding alternates investigated; and
f.) correspondence.

PART 2
DESCRIPTION OF PROFESSIONAL SERVICES

2-A. GENERAL

The Village of Massena is seeking proposals and costs for preliminary engineering services. The Village is also seeking cost estimates for design and construction services. Design and construction services estimates are being sought for information purposes and to assist the Village in evaluating proposals. This RFP is not for procurement of design/construction services.

The Village is also interested in total cost estimates for the proposed alternatives. The estimates should include the proposed cost of capital construction and the present value of the operation and maintenance for a 25-year period. The Village is interested in working closely with the successful proposer in the development of an affordable project. The proposers approach to performing the following tasks should take this into consideration and also be addressed in the work plan required in *Part 3, Section 3-C* of this RFP.

The Village may opt to utilize its own manpower for portions of the project. The Village may choose this option if opportunities exist to save significant dollars and if the Village has adequate manpower and expertise to accomplish the self help tasks. The tasks will be identified under the scope of work in *Part 2-B.m and 2-C.h.*

2-B. PRELIMINARY ENGINEERING PHASE

Services that proposers should incorporate in all proposals, as a minimum, include:

a.) reviewing all available information collected by the Village for the project. Advise Village contacts) if additional information is necessary for project development.

b.) reviewing current and projected land use information as it pertains to future wastewater usage.

c.) attending meetings, as necessary, with Village Board, sewer committee, and the public. Detail a public participation program which meets all federal, state, and local review and comment requirements.

d.) reviewing existing wastewater plant condition and performance. Inventory existing equipment and capabilities. Summarize performance capabilities as related to 1/1, influent and effluent characteristics, sludge disposal, and energy utilization and present SPDES limits.

e.) assessing upgrade and expansion needs for the collection system and the treatment plant. Develop several alternate approaches to the stated problems. Alternates should include the potential for modular expansion. At least one of the solutions must use proven conventional treatment processes. Solutions must be suited for cold climates and must include recommendations for sludge treatment, management, and disposal.

f.) evaluating treatment alternatives investigated and provide recommendations for the most appropriate and cost–effective wastewater collection and treatment technologies. Using a 25 year design period, state O&M cost impacts as a present worth for the various alternatives.

g.) working with the Village, DEC, EFC, and other governing authorities to obtain preliminary review approval of the best alternative. All design improvements must comply with the 1990 Edition of "Recommended Standards for Wastewater Facilities" now referred to as "GLUMRB" but formerly known as the "Ten State Standards for Wastewater Treatment."

h.) preparing a preliminary layout of the proposed facilities on 1 " =50' scale (or larger) current planimetric map including a hydraulic profile of the proposed facilities.

i.) revising preliminary layout per Village review.

j.) performing preliminary soil investigation.

k.) preparing a design report containing schematic layouts, sketches, design criteria, and associated calculations with appropriate exhibits.

l.) preparing cost estimates for project construction utilizing standard bidding practices.

m.) identifying self–help tasks that can be accomplished utilizing local forces. Prepare cost estimates for project construction utilizing self–help techniques. Cost estimates should identify labor, equipment, and material costs separately. Engineer will review task list with Village prior to incorporating into final design report.

n.) providing environmental review services. Engineer will complete SEQR/SERP and other necessary environmental assessment forms.

o.) modifying time schedule included in this RFP to reflect proposed construction schedule for the selected alternative.

p.) researching and evaluating funding options, including the EFC State Revolving Fund. As part of this phase, assist Village in the preparation of funding applications to the EFC and other identified funding sources.

q.) investigating alternate site locations for proposed treatment plant. Identify potential present worth cost for capital recovery and 25 years O&M.

r.) investigating PS#1 trunk main surcharge problems.

s.) investigating PS#1, 2, and 3 for condition. Make recommendations for possible rehabilitation .

t.) providing, as a minimum, the following preliminary design documents:
- final design report;
- revised cost estimate;
- preliminary outline drawings and hydraulic profile;
- detailed written project description;
- funding alternatives and best funding strategy;
- self–help analysis;
- revised time schedule;
- DEC and EFC preliminary project approval documentation.

2–C. FINAL DESIGN PHASE

The engineering scope and cost estimates requested for Final Design will be only preliminary estimates and largely dependent on the selected alternative. Estimates are being sought for services which, as a minimum, will include:

a.) complete topographic surveying as required for preparation of plans and specifications.

b.) performing staffing study based on final design of the wastewater plant. This study should offer recommendations of staffing levels and qualifications at the new plant consistent with plants with similar size and technology.

c.) preparing plans and specifications for the construction of the recommended wastewater treatment and collection facilities. Include MBE/WBE and affirmative action program requirements.

d.) obtaining soil borings as required, preparing soils analysis and incorporating in plans and specifications as appropriate.

e.) preparing contract documents in such a manner to meet all Federal and State requirements and securing approval of all regulatory agencies. Modifying documents as necessary to secure approvals.

f.) furnishing the necessary engineering data needed to apply for regulatory permits as required by local, state, or federal authorities.

g.) furnishing construction bidding and advertising services.

h.) preparing two (2) detailed cost estimates based upon final plans and specifications. One estimate utilizing standard bidding/contracting practices and one maximizing Self–Help techniques identified during the preliminary phase.

i.) preparing contract documents for Village and regulatory

agencies review and approvals.

j.) preparing a brief list of operation and maintenance tasks for the project, testing equipment needed, and testing procedures.

k.) preparing information and maps as needed for purchase and/or easements for acquisition of land for pump stations and treatment sites.

l.) providing the following final design documents (furnish up to 50 copies):

- Construction drawings,
- Construction specifications,
- Revised cost estimate,
- Construction contract documents,
- Proposed construction schedule,
- Progress schedule.

m.) setting up programs for MBE/WBE, equal opportunity, and affirmative action compliance.

2–D. CONSTRUCTION ASSISTANCE

Provide engineering scope and cost estimates for construction assistance for the construction phase of the project. Estimates for the following services, as a minimum, should include:

a.) assisting the Village in the construction phase of the project including project bidding, stakeout, scheduling, reviewing shop drawings, construction inspection and other services, as necessary.

b.) providing start–up assistance for the new plant and

equipment.

c.) preparing O&M manuals including a detailed section on staffing requirements to effectively operate and control the new plant.

d.) providing accurate "As Built" drawings of final construction with specific attention to piping layout, valve location, electric distribution and electric control schematics.

e.) furnishing a resident engineer on site at all times during construction.

PART 3
INFORMATION REQUIRED FROM PROPOSERS

To assist in the evaluation, all proposals must be submitted in the format as outlined below:

3-A. STATEMENT OF THE PROBLEM

Clearly state your understanding of the problem based on this RFP and the information that is available for review.

3-B. MANAGEMENT SUMMARY

Include a narrative description of the proposed effort and a list of the products that will be delivered. In developing this narrative, use the information included in this RFP as a starting point to ensure the Village's concerns are addressed. Engineers may go further and present options or alternatives, together with associated costs, where appropriate and where consistent with the Village's overall objective. The narrative should address any other concerns, issues, or technical requirements the Village should consider.

3-C. WORK PLAN

Describe in narrative form your technical plan for providing the services identified in **Part** 2. The Village acknowledges that the description of the plan for providing design and construction services may not be able to be as detailed as that for preliminary engineering because the project has yet to be identified. A description for these services should at least address the general scope of service and the manner in which it would be provided.

Modifications of the task descriptions are permitted; however reasons for changes should be fully explained. Indicate the number of man–hours you have allocated for each task. Specify which tasks can be completed by local officials and/or community volunteers. Cost for each task should be provided separately as specified in *Section 3–G.*

For each phase of the project (preliminary, design, and construction), present a task and subtask schedule. Submit a list of deliverables for each task. Deliverables include the minimum items indicated above in this RFP, and other reports, plans, meetings and specifications you deem necessary.

3–D. PRIOR EXPERIENCE

Submit prior experience and technical competence in designing and implementing the operation of community wastewater facilities. Wherever possible, projects should be of similar or larger size than this project. Include specialized experience in evaluation and design of wastewater treatment systems. This experience should include the complete scope of work necessary for the collection of data, determination of requirements, studies, system design, installation of equipment, testing and acceptance, and operation of wastewater systems. Also submit experience as related to procurement of funding (State Revolving Fund, etc.), working experience with the NYS DEC Region 6, and design of cold climates wastewater plants.

Experience should include work on self–help projects where force account work and other techniques using local resources have been employed.

Experience submitted should represent completed work by the project engineer to be assigned to the project. Studies or projects referred to should be identified and the name of the client provided, including the name, address, and phone number of the responsible official. For each project, also submit any identifiable self–help activities and resultant cost savings, total project cost, percentage of change order costs compared to total project cost, and date project completed.

3–E. PERSONNEL

Identify and include resumes of the executive and professional personnel that will be employed in the work. Show where these personnel will be physically located during the time they are engaged in the work. Include education and experience of the personnel in developing plans for wastewater facilities.

3–F. INSURANCE

Provide certificates demonstrating the following insurances:

    a.) Workmen's Compensation;
    b.) Professional Service Liability/Errors or Omission; and
    c.) Comprehensive General Liability.

3–G. COST AND PRICE ANALYSIS

The information requested in this section is required to support the reasonableness of your quotation. <u>This portion of the proposal must be bound and sealed separately from the remainder of the proposal.</u>

Tasks for the Preliminary Engineering, Final Design, and Construction Assistance phases must be separated with a total cost estimate for each.

The following cost categories must be identified for each phase:

a.) Manpower Costs: Itemize to show the following for each task:
- category, e.g. project manager, senior engineer, etc.;
- estimated hours;
- rate per hour; and
- total cost for each category and for all man power.

b.) Cost of Supplies and Materials: Itemize.

c.) Subcontract Costs: Itemize. This category includes all sub–contracts for services other than those included in a.) above.

d.) Transportation Costs: Show travel costs and per diem separately.

e.) Overhead.

f.) Profit.

g.) Total Cost.

-- END --

APPENDIX F

Adapted from FmHA Instruction 1942–A
(Guides 7 and 8)

PRELIMINARY ENGINEERING REPORT
WATER AND SEWERAGE

I. GENERAL. A Preliminary Engineering Report should clearly describe the owner's present situation, analyze alternatives, and propose a specific course of action, from an engineering perspective. The level of effort required to prepare the report and the depth of analysis within the report are proportional to the size and complexity of the proposed project. Farmers Home Administration projects must be modest in design, size and cost, and be constructed and operated in an environmentally responsible manner. The following should be used as a guide for the preparation of Preliminary Engineering Reports for FmHA financed water and wastewater systems.

II. PROJECT PLANNING AREA. Describe the area under consideration. The project planning area may be larger than the service area determined to be economically feasible. The description should include information on the following:

   A. Location. Maps, photographs, and sketches. These materials should indicate legal and natural boundaries, major obstacles, elevations, etc.

   B. Environmental Resources Present. Maps, photographs, studies and narrative. These materials should provide information on the location and significance of important land resources (farmland, rangeland, forestland, wetlands and 100/500 year floodplains, including stream crossings), historic sites, endangered species/critical habitats, etc., that must be considered in project planning.

C. **Growth Areas and Population Trends.** Specific areas of concentrated growth should be identified. Population projections for the project planning area and concentrated growth areas should be provided for the project design period. These projections should be based on historical records with justification from recognized sources.

III. **EXISTING FACILITIES.** Describe the existing facilities including at least the following information.

A. **Location Map.** Provide a schematic layout and general service area map (may be identified on project planning area maps).

B. **History.**

C. **Condition of Facilities.**

For Drinking Water. Describe present condition; suitability for continued use; adequacy of water supply, adequacy of current facilities; and, if any existing central facilities, the treatment, storage, and distribution capabilities. Also, describe compliance with Safe Drinking Water Act and applicable State requirements.

For Wastewater. Describe present condition; suitability for continued use; adequacy of current facilities; and, if any existing central facilities, the treatment, storage, and disposal capabilities. Also, describe compliance with Clean Water Water Act and applicable State requirements.

D. **Financial Status of any Operating Central Facilities.** Provide information regarding rate schedules, annual operating and information (O&M) cost, tabulation of users by monthly usage categories and revenue received for last three fiscal years. Give status of existing debts and required reserve accounts.

IV. NEED FOR PROJECT. Describe the needs in the following order of priority:

   A. **Health and Safety**. Describe concerns and include relevant regulations and correspondence from/to Federal and State regulatory agencies.

   B. **System O&M**. Describe the concerns and indicate those with the greatest impact. Investigate water loss (for drinking water systems) and infiltration and inflow (for wastewater systems), management adequacy, inefficient designs, and problem elimination prior to adding additional capacity.

   C. **Growth**. Describe the reasonable growth capacity that is necessary to meet needs during the planning period. Facilities proposed to be constructed to meet future growth needs should generally be supported by additional revenues. Consideration should be given to designing for phased capacity increases. Provide number of new customers committed to this project.

V. ALTERNATIVES CONSIDERED. This section should contain a description of the reasonable alternatives that were considered in planning a solution to meet the identified need. The description should include the following information on each alternative:

   A. **Description**. Describe the facilities associated with the alternative.

   Drinking Water. Describe all feasible water supply

sources and provide comparison of such sources. Also, describe treatment, storage and distribution facilities.

Wastewater. Describe all feasible wastewater treatment technologies and provide comparison of such. Also, describe collection facilities. A feasible alternative may be a combination of central facilities and management of on–site facilities or only the later.

B. **Design Criteria**. State the design parameters used for evaluation purposes. These parameters must follow the criteria established in FmHA Instruction 1942–A.

C. **Map**. Schematic layout.

D. **Environmental Impacts**. Do not duplicate the information in the applicant's submittal of environmental information. Describe unique direct and indirect impacts on floodplains, wetlands, other important land resources, endangered species, historical and archaeological properties, etc., as they relate to a specific alternative. FmHA must conduct an environmental assessment prior to project approval.

E. **Land Requirements**. Identify sites and easements required. Further specify whether these properties are currently owned, to be acquired or leased.

F. **Construction Problems**. Discuss concerns such as subsurface rock, high water table, limited access or other conditions which may affect cost of construction or operation of facility.

G. **Cost Estimates**.
    1. Construction.

2. Non–Construction and Other Projects.
3. Annual Operation and Maintenance.
4. Present Worth, based on Federal discount rates.

H. Advantages/Disadvantages. Describe the specific alternatives ability to meet the owner's needs within its financial and operational resources, comply with regulatory requirements, compatibility with existing comprehensive area–wide development plans, and satisfy public and environmental concerns. A matrix rating system could be useful in displaying the information.

VI. PROPOSED PROJECT (RECOMMENDED ALTERNATIVE).
This section should contain a fully developed description of the proposed project based on the preliminary description under the evaluation of alternatives. At least the following information should be included:

A. Project Design.

1. **Drinking Water.**
   a. **Water Supply.** Include requirements for quality and quantity. Describe recommended source, including site.
   b. **Treatment.** Describe process in detail and identify location of plant and site of any process discharged.
   c. **Storage.** Identify size, type and site location.
   d. **Pumping Stations.** Identify size, type site location and any special power requirements.
   e. **Distribution Layout.** Identify general location of line improvements: lengths, sizes and key components.
   f. **Hydraulic Calculations**. This information should provide sufficient detail in a tabular format to de-

termine compliance with FmHA design requirements. Automation tools may be used by the engineer. The submittal should include a map with a list of nodes and pipes and the associated characteristics, such as elevation of node, pipe diameter, pipe segment length, reservoir elevation, domestic and industrial water demands, fire flow, etc.

2. **Wastewater.**

   a. **Treatment.** Describe process in detail and identify location of plant and site of any discharges.
   b. **Pumping Stations.** Identify size, type, site location and any special power requirements.
   c. **Collection System Layout.** Identify general location of line improvements: lengths, sizes and key components.
   d. **Hydraulic Calculations**. This information shouldprovide sufficient detail in a tabular format to determine compliance with FmHA design requirements. Automation tools may be used by the engineer. The submittal should include a map with a list of manholes and pipes and the associated characteristics, such as elevation of inverts, pipe diameter, pipe segment length, etc.

B. **Cost Estimate**. Provide an itemized estimate of the project cost based on the anticipated period of construction. Include development and construction, land and rights, legal, engineering, interest, equipment, contingencies, refinancing, and other costs associated with the proposed project. (For projects containing both water and waste disposal systems, provide a separate cost estimate for each system.)

C. Annual Operating Budget.

1. **Income.** Provide a rate schedule. Project income realistically, based on user billings, drinking water or wastewater treatment contracts, and other sources of incomes. In the absence of other reliable information, for budget purposes, base water use (or in the case of wastewater, wastewater generation) on 60 gallons per capita per day, or 150 gallons per residential–sized connection per day, or 4,500 gallons per residential–sized connection per month. When large agricultural or commercial users are projected, the report should include facts to substantiate such projections and evaluate the impact of such users on the economic viability of the project. The number of users should be based on equivalent dwelling units, which is the level of service provided to a typical rural residential dwelling.

2. **Operations and Maintenance Costs.** Project costs realistically. In the absence of other reliable data, base on actual costs of other existing facilities of similar size and complexity. Include facts in the report to substantiate operation and maintenance cost estimates. Include salaries, wages, taxes, accounting and auditing fees, legal fees, interest, utilities, gasoline, oil and fuel, insurance, repairs and maintenance, supplies, chemicals, office supplies and printing and miscellaneous.

3. **Capital Improvements.**
   For Drinking Water. If purchasing water or if water is being treated by other, these costs should be included in O&M costs.

4. **Debt repayments**. Describe existing and proposed project financing from all sources. All estimates of FmHA funding should be based on loans, not grants. FmHA will evaluate the proposed project for the possible inclusion of FmHA grant funds.

5. **Reserve.** Unless otherwise required by State statute, establish at one–tenth (1/10) of annual debt repayment requirement.

VII. CONCLUSIONS AND RECOMMENDATIONS. Provide any additional findings and recommendations that should be considered in development of the project. This may include recommendations for special studies, highlight the need for special coordination, a recommended plan of action to expedite project development, etc.

APPENDIX G

REFERENCES

Here is a list of references mentioned in this book, along with a number of additional publications that you might find useful. This list is by no means a comprehensive bibliography. Many other valuable publications can be obtained by contacting the organizations described on pages 177-179.

Note: For your convenience, we have included telephone numbers for the publishers of these documents. Unless otherwise noted, EPA materials can be ordered from the National Small Flows Clearinghouse and/or the National Drinking Water Clearinghouse (800-624-8301). For publications of STEP and The Rensselaerville Institute, call us at 518-797-3783.

GENERAL

A Brighter Future for Rural America? Strategies for Communities and States. DeWitt John, Sandra S. Batie, and Kim Norris. National Governor's Association Center for Policy Research, 1988. Tel. 202-624-5300. Outlines the competitive challenge facing rural America and reports on signs of hope for the rural economy.

Civic Environmentalism: Alternatives to Regulation in States and Communities. DeWitt John. CQ Press, 1994. Tel. 202-887-8500. Advocates a new style of environmental politics and policy based on decentralized, bottom-up initiatives that use new tools to address environmental problems.

Clues to Rural Community Survival, 6th Edition. Heartland Center for Leadership Development, 1990. Tel. 402-474-7667. Lists 20 factors associated with successful local efforts to survive difficult conditions and discusses case studies that illustrate those factors.

Communities in the Lead: The Northwest Rural Development Sourcebook. Harold S. Fossum. Northwest Policy Center, Univ. of Washington, 1993. Tel. 206-543-7900. Describes resources, tools, and strategies for community leaders involved in local revitalization efforts.

Environmental Planning for Small Communities: A Guide for Local Decision-Makers. U.S. EPA Publication No. 625/R-94/009, 1994. Tel. 513-569-7562. Offers detailed, step-by-step advice on how to develop a community environmental plan. Useful appendices include a relevant case study, a discussion of regulations that affect small communities, suggestions for risk assessment, and lists of resources.

Tapping Your Own Resources: A Decision-Maker's Guide for Small Town Drinking Water. National Association of Towns and Townships, 1993. Tel. 202-737-5200. A brief primer for local officials and others involved in the management of small water systems. Gives an overview of responsibilities and problems facing managers, and suggests strategies for meeting these challenges.

Treat It Right: A Local Official's Guide to Small Town Wastewater Treatment. National Association of Towns and Townships, 1989. Tel. 202-737-5200. A general introduction to wastewater treatment for community leaders. Covers the local government's role, planning for a project, technical options, financing, and management responsibilities.

What Does It Cost to Save Money? The View from the States. Harold S. Williams. The Rensselaerville Institute, 1994. A look at the costs and benefits to states of participating in STEP. Features a hypothetical illustration that is lighthearted but politically realistic.

INTERLOCAL COOPERATION & PARTNERSHIPS

Intergovernmental Cooperation. NYS Department of State, 1989. Tel. 518-473-3355. Includes discussion of legal requirements for intermunicipal partnerships in New York State.

Partnerships for Small System Viability: A Process Guide and Three Case Studies. Small Towns Environment Program. The Rensselaerville Institute, 1994. Offers guidelines for establishing interlocal partnerships and presents cases of successful cooperative initiatives.

Self-Help Partnerships for Small Communities. Small Towns Environment Program. The Rensselaerville Institute, 1994. How interlocal partnerships can enable small communities to address water and wastewater problems that are too expensive to resolve through individual action.

"Tug Hill, New York: Progress Through Cooperation in a Rural Region." Benjamin P. Coe. *National Civic Review,* Fall-Winter 1992, pp. 449-465. Tel. 303-571-4343. Presents the case of a rural region where state-local cooperation and resource pooling between local governments has raised indigenous problem-solving capacity.

"Working Together: You've Got a Lot to Gain." *Community Water Bulletin*, special edition. Community Resource Group/Southern RCAP, undated. Tel. 501-756-2900. This special issue of CRG's newsletter discusses mutual aid among small systems.

PRIVATIZATION

Public-Private Partnerships for Environmental Facilities: A Self-Help Guide for Local Governments. U.S. EPA Publication No. 20M-2003, July 1991. Discusses the main types of public-private partnerships, steps for building one, financing, and contract considerations. Includes an appendix listing useful organizations and publications.

Rural Water/Wastewater Study, Vol. I: Background, Terminology and Recommendations. Rural Electric Research/ National Rural Electric Cooperative Association, 1992. Tel. 202-857-9500. Examines how rural electric cooperatives can use their technical, financial and administrative capacity to assist community water and wastewater systems.

"When Privatization Makes Sense." William Gehr and Michael Brown. *BioCycle,* July 1992, pp. 66-69. Tel. 215-967-4135. Presents factors to consider in structuring a privatization arrangement. Focus is on waste management projects, but most points are also relevant to privatization of other services/facilities.

PROJECT FINANCING

Creating Partnerships: A Local Action Guide for Rural Development. Virginia Water Project/Southeast Rural Community Assistance Project, Fall 1988. Tel. 703-345-1184. A concise working guide to obtaining private sector aid for rural water and wastewater projects.

Financing Water and Wastewater Disposal Systems in Rural Areas: A Working guide for State Program Coordination. Council of State Community Development Agencies, May 1994. Tel. 202-393-6435. This guide is primarily for state water and wastewater officials, but has information on three major funding programs (Community Development Block Grants, the Rural Development Administration [now the Rural Utilities Service] and state revolving funds) that others may find useful as well.

Handbook of Project Finance for Water and Wastewater Systems. Michael Curley. CRC Press, Boca Raton, FL, 1993. Tel. 800-272-7737. Explains in a detailed but understandable way the tools and concepts needed for financial management of water and wastewater projects. Particularly useful for small system board members who are not trained in finance.

Innovative Grassroots Financing: A Small Town guide to Raising Funds and Cutting Costs. National Association of Towns and Townships, 1990. Tel. 202-737-5200. Presents a variety of strategies for financing programs and services in small communities, including special events, mini-bonds, user fees, public-private partnerships, and local production.

PROJECT MANAGEMENT

Estimating for the General Contractor. Paul J. Cook. R.S. Means Company, 1985. Tel. 617-585-7880. A practical reference book calculating labor, materials, etc. needed for a project. Very useful for making cost estimates.

Flow Chart for Project Managers. Small Towns Environment Program. The Rensselaerville Institute, 1994. Shows the process of a typical self-help water or wastewater project. Identifies in sequence important tasks as well as the key actors who perform them.

Project Task Matrix. Small Towns Environment Program. The Rensselaerville Institute, 1989. A set of worksheets for specifying self-help vs. contracted components of a project. Lists action steps and provides spaces for indicating who (the community or an outside agent) will perform a given task, by what date, and at what cost.

REGULATIONS

A Guide to Federal Environmental Requirements for Small Governments. U.S. EPA Publication No. 270/K-93/001, September 1993. A reference handbook on the major environmental regulations with which small jurisdictions must comply. Covers drinking water, wastewater, wetlands, solid waste, toxics and air quality.

The Small System Guide to the Safe Drinking Water Act, Second Edition. Community Resource Group/Southern RCAP, August 1993. Tel. 501-756-2900. Examines small system responsibilities under the SDWA, including the Surface Water Treatment Rule, monitoring and testing for contaminants, public notification requirements, etc.

SMALL SYSTEM O&M AND VIABILITY

The Board Guide to Small System Policies. Community Resource Group/Southern RCAP, 1993. Tel. 501-756-2900. One of CRG's excellent series designed for volunteers, local officials and others who serve on the boards of directors of small water systems. Explains why policies are necessary and what makes for a good policy, and gives a comprehensive set of sample policies.

Community Water Bulletin series on small system viability. Community Resource Group/Southern RCAP, 1993. Tel. 501-756-2900. Relevant editions are:
> "What is Small System Viability'?" (No. 75)
> "Upcoming Regs: Can You Remain Viable?" (No. 76)
> "More Upcoming Regs: Assess Your Liability Now" (No. 77)
> "Assessing Your Water Supply" (No. 78)
> "Assessing the Condition of System Components" (No. 79)
> "Administration: Will Yours Stand the Test of Time?" (No. 80)
> "Finances for the Future" (No. 81)

The Small System Guide to Developing and Setting Water Rates. Community Resource Group/Southern RCAP, 1993. Tel. 501-756-2900. Covers how to evaluate a system's financial health, the need for metering, how to use customer usage information, and methods for determining sound and equitable rates.

The Small System Guide to Risk Management and Safety. Community Resource Group/Southern RCAP, 1993. Tel. 501-756-2900. Explains board liability and suggests policies and procedures for handling chlorine, traffic control, confined space/excavation work, electrical hazards and emergency planning.

TECHNOLOGIES

Drinking Water Treatment for Small Communities: A Focus on EPA's Research. U.S. EPA Publication No. 640/K-94/003, May 1994. This pamphlet summarizes EPA's drinking water program and gives an overview of risk, treatment and current research for various categories of drinking water contaminants.

Small Community Water and Wastewater Treatment. U.S. EPA Publication No. 625/R-92/010, 1992. An overview of technical options appropriate for small settlements.

Small Wastewater Systems: Alternative Systems for Small Communities and Rural Areas. U.S. EPA Publication No. 830/F-92/001, May 1992. Diagrams and brief explanations of various wastewater collection/disposal systems, especially onsite and cluster technologies.

Part Six

Index

A

"as-built" drawings, 175
affordability, 2, 3, 6, 27, 30, 36-38, 50, 81, 82, 85, 106, 110, 112, 119-124, 164
American Water Works Association, 142, 178
assessment, 40-44
Association of State and Interstate Water Pollution Control Administrators, 189
attorneys, 108-109, 117, 128, 148, 16

B

banks, *See* financing
Bell, OK, 4, 87
benefits, 60, 62, 98, 140-141
Better Business Bureau, 145-146
bidding, 62, 73-78, 145, 158
billing, 118, 168, 173
bonds, *See* financing
Boonville, NY, 48
borrowing equipment, 69-71
business, 23-24, 73, 155

C

capital 24, 41-42, 122-144
 See also financing
Cherokee Nation, 4, 87
child care, 96-97
Civil Service, 61, 170-171
Clean Water Act (CWA), 134-135, 165, 187-189, 191-192
collaboration, 17, 18-20
colleges and universities, 143
commitment, 29, 89, 96, 131-132
communication, 17, 90, 102-105, 156
community, 1, 2, 3, 4, 5, 27, 29, 198-200
 control, 111, 113, 119-121, 122, 141, 159, 161, 167, 170, 172
 definition, 195-198
 mobilizing, 28-29, 87-99, 141-144
 potential, 23-27, 31
 readiness, 28-30, 31
 values, 200-203
Community Development Block Grant (CDBG), 137-138
compacts, 15-17, 18
comparison shopping, 73, 122
competition, 97, 110, 138
compliance, regulatory, 12, 41, 42, 121, 123, 135, 136, 162, 164-165, 187-194
Connelly Springs, NC, 87-89
conservation, 45-47
construction, 90, 92-95, 115, 116
 equipment, 69-71, 72-78
 inspection, 36, 59, 61,

85, 116, 148
local workers, 59-68
manager, 167-170
materials, 72-78
schedule, 167-170
supervision, 90
contracts, 74, 77, 109, 111, 114, 116
contractors, 36, 37, 49, 59, 62, 72, 73, 80, 85, 100-101, 145-148, 158
Corbett, NY, 4, 45
corporations, 131-132, 142
costs
    administrative, 36, 90, 100, 140
    capital, 41, 49, 80
    construction, 35-38, 41, 125, 145
    energy, 46, 133
    engineering, 36-37, 111-112, 163
    estimating, 73, 91, 110, 120, 163-164
    insurance, 62, 91, 150, 151, 153-160
    interest, 120-124, 127
    labor, 59-68
    legal, 77
    of grants, 119, 194
    operation and maintenance, 41, 46, 49, 50, 133, 145, 166, 174, 189
    project case examples, 40-43, 44, 46, 66, 71, 80-84, 88, 123, 133
    retail, 36-39, 66
    savings, 2-4, 12-13, 27, 35-39, 40-43
    transportation, 74, 76
Council of State Community Development Agencies (COSCDA), 135-136, 138

D

Davis-Bacon Act, 62-63
Delmar, MD/DE, 46
design criteria, 44, 47, 49, 90, 91-92, 100, 110-111, 113, 114, 116, 140, 145, 163, 168
direct purchase, 72-78
Dolgeville, NY, 123
drinking water
    information, 165, Appendix G
    potential contaminants of, 191
    project examples, 40-42, 46, 50, 66, 71, 88, 123
    revolving loan funds, 135, 187, 192
    systems, 190-192
    technology, 11, 40-43, 50, 123

E

easements 1, 27, 100, 169
economic development, 136, 137, 194, 198-200

Edison and Blanchard, WA, 44, 104
enabling role, *See* government
energy efficiency, 46, 133
enforcement, 3, 12, 14, 31, 192, 194
engineers/engineering
    contracts, 36, 37, 85, 112, 114, 116, 117, 118
    estimate, 38, 73, 121, 164
    inspection, 36, 59, 85, 94
    overdesign, 49
    oversight, 41, 93, 168
    reference books, 163
    reports, 38, 49, 100, 111, 114, 117, 161, 162-164, 166
    selection, 110-118
    technical options, 40-43, 49-50, 123, 145, 163
entrepreneurs, 23-26
Environmental Protection Agency, U.S. (EPA), 2, 6, 44, 50, 59, 64, 139, 141, 142, 187, 193
    Construction Grants Program, 188
    Needs Survey, 188-189, 225
Equal Employment Opportunity Act, 63

F

Farmers Home Administration, *See* Rural Economic and Community Development
feasibility studies, 119
Ferguson, Doug, 73, 180
financing, 35-38, 119-144
    banks, 38, 70, 122-128, 133, 148
    bond anticipation notes (BANs), 125
    bond banks, 128
    bonds, 125-128, 135
    cash flow, 122
    capital notes, 126
    corporations, 131-132
    developer, 138, 139
    foundations, 128-131
    from individuals, 132
    interest rates, 120, 122-124
    leveraging, 65
    loans, 14, 122-128, 134-138, 187-188, 192
    notes, 125-126
    repayment capacity, 119-121
force accounting, *See* labor
Ford Foundation, The, 2, 6
foundations, *See* financing

G

General Duty Doctrine, 149
government, role of, 11-20, 194

grants, 119, 129-132, 134-138
guarantees
    from engineers, 116
    from vendors, 73, 74,141
    on self-installed material, 60

H
Heuvelton, NY, 133
Housing and Urban Development, U.S. Department of (HUD), 7, 137-138

I
import substitution, 35, 199
in-kind contributions, 141-144 *See also* financing
infrastructure needs, 187-194
insurance, 149-160
    automobile, 150-151
    builders' risk, 153
    comprehensive, 152
    deductibles, 157
    for volunteers, 156
    general liability, 151-152
    installation floaters, 153-154
    municipal bond insurance (MBI), 127-128
    risk, 25-26, 31, 38, 91, 125, 128, 140, 141, 149-150, 160, 192
    self-insurance, 151, 160
intergovernmental agreements, 79-84, 88, 194
inventory, 72, 95

L
labor, 3, 59-68, 148 *See also* volunteers
    force accounting, 59-66
    municipal, 27, 35, 59-66, 85
    prevailing wage rates, 59, 62, 75, 148
    prisoners as workers, 68
    public assistance recipients, 67
    sources of, 59-68
lawyers, *See* attorneys
leaders, 201
leakage, 46, 73
learning, 20, 25, 94, 175
leasing, 70
lessons learned 27, 81, 175 *See also* learning
liability *See* insurance
loans *See* financing
local government *See* government
lump-sum payment by users, 138

M
mandates, 4, 191
maintenance schedule, 174
Marshall, NC, 66
Massena, NY, 113
materials

considerations, 90-91, 101
contractor's markup for, 36
direct purchase, 35-37, 72-78
engineer's specifications for, 110-111, 145
liability, 73, 150-151,153-154
recycled, 72-73
responsibility for, 73, 169
"scratch and dent", 72
state laws and bidding, 73-78
Means Guides, 163
media, 102-103, 176
Median Household Income (MHI), 136-137
member items, 139
meters, 45-46
milestones, 35, 102
Minority Business Enterprises (MBE), 63-64
money *See* financing
Monthly Bill to Clients, engineer's, 118
motivation
    compacts and, 15-17
    morale, 89, 96-99
    project manager and, 101
    sparkplugs and, 26
    techniques for, 97-98
    volunteer projects and, 46, 88, 89

waiver of liability and, 91

N

National Drinking Water Clearinghouse, 165, 178
National Small Flows Clearinghouse, 165, 178
New Berlin, NY, 50
New Jerusalem, Arkansas, 1, 71, 87, 129, 142
nonprofits *See* organizations

O

onsite systems *See* wastewater systems
operation and maintenance (O&M), 30, 81-84, 88, 123, 139, 166, 189
    contracting, 139
    preparation for, 173-174
organizations, 18, 46, 67, 69, 105, 129, 132, 134, 142, 143, 177-179

P

packaging, 85, 136
partnerships, *See* intergovernmental agreements *and* public-private partnerships
Perley, Diane, 38, 181
personnel *See also* labor
    coordination of, 169
    hiring, 36-37, 67, 107, 117, 123, 133, 153, 170
    operations, 173-174

productivity and, 97
technicians, 93, 142
potential, 19, 23-27, 161
privatization *See* public-private partnerships
procurement, 35-38, 59, 72-78, 80-84, 109, 140, 142
project(s), 40-42, 44, 46, 50, 66, 71, 80-84, 87, 88, 100, 123, 133
    manager, 90, 123, 167-170
    Task Matrix, 110
publications *See* Appendix G
publicity *See* media
public-private partnerships, 139-141, 194
purchasing *See* procurement

Q
quality control, 94, 115

R
rate structure, 82, 83, 173
readiness, 19, 23, 28-31
record-keeping, 59, 90
references, 109, 111, 115, 145
regulations, 64, 67, 136, 165, 170-172, 188-194
See also compliance, regulatory
reinventing, 13, 25
Rensselaerville Institute, The, 2, 13, 16, 18, 87, 110
repayment capacity *See* affordability *and* financing restructuring, 192 *See also* intergovernmental agreements
risk, 25-26 *See also* insurance
Rural Community Assistance Program, 142, 177-178
Rural Development Administration *See* Rural Economic and Community Development
Rural Economic and Community Development (RECD), 45, 66, 100, 117, 136-138, 196
Rural Water Association, 142, 177

S
Safe Drinking Water Act (SDWA), 165, 187, 189-193
Safe Drinking Water Hotline, 165
safety, 95, 145, 150
sanitary district, 75, 80, 82, 83
"scratch and dent" materials *See* materials
SDWA *See* Safe Drinking Water Act
self-help, definition of, 3-4
Self-Help Support System, NY, 4, 5, 17, 40-42, 50,

133, 180-181
septic systems *See* wastewater systems
Siddon, Steve, 113-14, Appendix E
skills, 27, 61, 71, 93-94, 99, 100, 108-109, 112, 145, 147
small systems, 79, 178, 191-193
Small Towns Environment Program (STEP), 4-5, 13, 16-20, 23, 30, 35, 44, 46, 66, 71, 80, 84, 86, 88, 94, 98, 110, 129, 156, 164, 175
Smyrna, NY, 100
sparkplugs, 19, 23-27, 99
state government *See* government *and* compliance
state revolving funds (SRFs)
   See also financing
   characteristics, 134-136
   drinking water, 14, 135, 192
   wastewater, 14, 134-136, 187-190
STEP *See* Small Towns Environment Program
Surface Water Treatment Rule, 42, 190
sweat equity, 3

T

tax credits, 140
technology *See* drinking water and/or wastewater systems
technical assistance, 14, 18, 140, 141, 177-183, 192
training, 18, 48, 93, 94, 101, 109, 174, 177-178, 193

U

unincorporated areas, 29
user fees, 38, 41, 45, 79, 82, 120, 121, 122, 124, 174, 189, 191 *See also* drinking water *and* wastewater

V

viability, 192
volunteers, 3, 44, 67, 87-99, 100, 116, 151

W

warranties, 60, 73, 146
Washington Co. (MD) Sanitary District, 75
wastewater, 3, 14, 34, 136
   federal regulation, 187-189
   funding sources, 134-138
   onsite/septic systems, 15, 44, 47
   project examples, 44, 51-58, 80-84, 100, 133, Appendix B
   sewer service, cost of, 188-189
   technical assistance,

177-183
   technologies, 45, 50-58, 142
   user fees, 121, 126, 138
water quality, 50, 66, 174, 187-194
   projects for protection of, 44, 66
   testing, arranging for, 174
Water Environment Federation, 142, 177
Western Maryland Cooperative Utilities Venture (WMCUV), 80-84, Appendix B
women, 88, 94, 142, 143
Women's Business Enterprises (WBE), 63-64
work crews, 54-65, 97
Workers' Compensation, 67, 91, 160